The Effective Ecologist

Succeed in the Office Environment

Neil Middleton

Illustrations by Joan Punteney

Pelagic Publishing | www.pelagicpublishing.com

GULL LAKE

Published by Pelagic Publishing
www.pelagicpublishing.com
PO Box 725, Exeter EX1 9QU, UK

The Effective Ecologist: Succeed in the Office Environment

ISBN 978-1-78427-083-4 (Pbk)
ISBN 978-1-78427-084-1 (ePub)
ISBN 978-1-78427-085-8 (Mobi)
ISBN 978-1-78427-086-5 (PDF)

Excel is a registered trademark of the Microsoft Corporation.
For more information visit www.microsoft.com.

British Library Cataloguing in Publication Data
A catalogue record for this book is available from the British
Library.

Cover image by Joan Punteney
Typesetting by Saxon Graphics Ltd, Derby

ABOUT THE AUTHOR

Neil Middleton has over 35 years' experience within the service, financial and ecological sectors, performing a variety of leadership, managerial, technical, customer service, people development, consultancy, marketing and financial roles. He is the managing director of two companies operating within the UK: Echoes Ecology Ltd, an ecological consultancy he established in Scotland in 2006, and Time For Bespoke Solutions Ltd, a company that provides management consultancy and people development solutions. Neil is also an accomplished trainer across a wide range of business and ecology-related subjects, having developed and delivered well over 200 training events to date. He has a constant appetite for self-development, as well as seeking to develop those around him. Hence the inspiration behind this book, in which he shares his experiences, thoughts and ideas as to how best you can perform successfully in your role.

Neil is available to provide people development and management consultancy services to ecological consultancies and other service-sector businesses throughout the UK.

neil.middleton@timefor.co.uk
www.echoesecology.co.uk and **www.timefor.co.uk**

Also available by the same author:
Social Calls of the Bats of Britain and Ireland. (2014): Neil Middleton, Andrew Froud and Keith French. Pelagic Publishing, Exeter. ISBN: 978-1-907807-97-8.

ABOUT THE ILLUSTRATOR

Joan Punteney studied illustration and graphic design at Edinburgh College of Art. She works as a freelance artist on a wide variety of projects and subjects. Her main work, and passion, has always been painting animals, and she has undertaken countless commissions for lifelike paintings of horses, pets and wildlife. Joan is also an excellent cartoonist, as demonstrated by the work produced for this book. Joan is available to produce commissioned high-quality artwork.

jlpunteney@yahoo.co.uk

CONTENTS

Preface   vii
Acknowledgements   xi

Chapter 1   Being effective   1

Chapter 2   What's the job?   13

Chapter 3   Positive behaviours   27

Chapter 4   Communication skills   49

Chapter 5   Organisational skills   83

Chapter 6   Meetings, meetings, more meetings   113

Chapter 7   Project management   129

Chapter 8   Reporting   155

References   179
Appendix 1   Feedback: get rich quick   180
Appendix 2   Effective allocation of tasks   182
Appendix 3   Effective brainstorming   186
Glossary   188
Index   198

me that *'Every day, in every way, you're getting better.'* Who would have thought that would still prove to be such a source of inspiration, some 35 years later? And then there were my colleagues at Norwich Union PLC. Too many to mention here, but in particular I am grateful to Ian Brodie, Isobell Carroll, Ian Girdwood, Tracey Henderson, Mike King, Scott McLean, Peter Millington, Steve Molyneux, Bill Petrie, Ian Potter, Janice Rodgers, John Shearer, Gordon Smith, Tim Webb and Phil Worthington. Although all of this seems so far away from the world of ecology, so much of what I have learned and achieved during my own professional journey is down to these people.

Collectively, all the people named above have taught me that the day you stop being receptive to new ideas, new challenges and self-development is the day you may as well just pack up everything, switch off the lights and wait for ... Wait for what, exactly? In professional life worthwhile things rarely happen unless you work hard and make them happen. You may be lucky, you may be superhuman, you may be the exception to the rule. As for the rest of us, if we really want it badly enough we have to work darn hard. Fortunately, all the way through my professional life, I have been lucky enough to work alongside and be inspired by people who know this.

ACKNOWLEDGEMENTS

It would have been impossible to write this book without a huge amount of assistance and support, not only while writing it, but also during my entire learning and working life.

First of all I would like to thank a number of people who have contributed to the production of this book: Joan Punteney, for the excellent front cover and the equally brilliant illustrations accompanying each chapter; Aileen Hendry, my partner in life, as well as my constant reviewer, for her knowledge, experience, feedback (lots of feedback, in fact!) and greatly needed proofreading skills; Laura Carter-Davis (Team Manager, Echoes Ecology Ltd) for her support and critical appraisal of the material, and for her proofreading; Hugh Brazier, for his excellent copy-editing skills and valued input; and Nigel Massen of Pelagic Publishing, for his faith that I had something worth saying and for his continued support and professional expertise.

I am grateful to Steve Jackson-Matthews (Head of Ecology and Director, Land Use Consultants Ltd), Reuben Singleton (Director, Tweed Ecology Ltd) and David Darrell-Lambert (Director, Bird Brain UK Ltd) for their valued opinions relating to the subjects covered in this book, and for allowing me to quote some of their wise words.

Special thanks are also due to Andrew Froud and Keith French, my co-authors on *Social Calls of the Bats of Britain and Ireland*, for their friendship and inspiration over many years.

For support and encouragement throughout the writing process, and at certain key moments, my gratitude goes to Heather Campbell, Paul Carter-Davis, Audrey Middleton, Emily Middleton and Sophie Punteney.

Within Echoes Ecology Ltd I have been blessed, through good planning and a small dose of luck, to have an excellent team, and in particular during the writing of this book thanks are due to Laura Carter-Davis, Heather Ream, Rhiannon Hatfield, Elaine Anderson, Aaron Middleton, Colin Everett, Laura Spence, Mingaile Zebaite and Emily Platt for being so good at what they do best: looking after our customers and working so effectively together.

I would finally like to acknowledge some key mentors from my earlier life in business. The management team at the Gloucester Hotel (Aberdeen) gave me my first lessons in customer service, and the head waitress, Dorothy, constantly reminded

CASE-STUDY CHARACTERS

Other than where I refer to real events – in which case nothing that can identify any individual is provided – all other characters mentioned by name or appearing in this work are fictional. Any resemblance to real persons, living or dead, is purely coincidental. Some of these fictional characters appear in a number of the case studies and examples. The same characters are used throughout, as follows:

Mr Mitchell (Bill)	Client (Smith & Co.)
Michael	Team manager
Jane	Senior ecologist
Robert	Ecologist 1 (good performer)
John	Ecologist 2 (poor performer)
Tom	Ecologist 3 (from a different office within the same company)

GLOSSARY

A glossary of regularly used business-related words and expressions appears at the end of the book. It is not a conventional glossary, in that it is not limited to defining words that occur in the text, but includes various other terms that you may encounter during your business-related activities. I hope that you find it useful.

So, there I was one day, not unusually for a man quickly approaching his grumpy years, wondering why so little has been done to help ecologists with this aspect of their working role. I quickly moved on to lambasting those who had never done anything about it. Now, I am not quite sure who the 'those' I was thinking about actually were, because to be fair very few of us, at least in our sector, are experts in business-related human behaviour. Strange, if we bear in mind that many of us know a considerable amount about animal behaviour. And of course, humankind is very much like the other creatures many of us study. Usually when we do or say something, or react in a certain way, it is for a reason or in response to our environment, including the behaviours of those around us.

Human behaviour is a massive subject, which I am clearly in no way qualified to comment upon. I do, however, have some relevant experience within the world of business, management, team building and the like, and I have spent a huge amount of time working in office environments (some would say too much!). With that in mind, here I am endeavouring to contribute towards improving business-related skills within the ecology sector. I hope this book will help you see your world through different eyes and give useful perspectives on how to manage your way through the potentially hazardous office habitat. On the face of it the office looks comfortable and safe, but on a bad day it can make the career you are so passionate about seem very challenging.

This book will cover lots of ideas for you to consider in order to make the office-related aspects of your performance less stressful and more effective. With these additional skills your chances of being an asset to your team and your employer will be enhanced, as will your ability to forge a successful career in your chosen profession.

The approaches described herein are by no means the only ones that can be adopted in a given scenario, but at the very least they give you more in your toolbox to help you to navigate effectively through your day.

Bear in mind that small changes to your behaviours can have a huge positive impact upon your performance and how you are perceived within your working environment. If, on the other hand, you change nothing, then nothing will change. It is a fine line between mediocrity and brilliance. Brilliance is not that difficult to achieve. It's all about adopting the correct approach to your working environment, and delivering what's expected of you every time; no excuses. Be good at the simple stuff, and everyone around you will perceive you as brilliant.

At first glance this book might appear to be aimed purely at those working as an ecologist, at whatever level. However, if you are in a more senior position (such as a team manager or principal ecologist), or working in a human resources (HR) or training role, there is a lot in here that may give you food for thought as to how to approach different scenarios, or how to encourage people around you to develop. At the very least you may wish to steer your team members towards reading this book.

PREFACE

Those working in ecology are technically a highly competent group of people. Almost every ecologist, at every level, has either completed a degree associated in some way with their role, or has amassed a huge amount of experience in their core subject matter, or has even managed to achieve both. Many of us, as we grew up, watched birds, studied bats, recorded insects and identified plants, all as part of our leisure time. To progress from being a keen amateur to gaining a professional position in a field that we are passionate about is a dream come true. In fact, in many industry sectors you will struggle to find a workforce anywhere near as emotionally attached at the same personal, non-working level. As the working week draws to a close I might ask a colleague, *'So what are your plans this weekend?'*, and the number of times I have been given an answer such as *'I am going to do a survey for the local badger group'* is a huge credit to the drive and commitment many show. I cannot imagine many other professions where people spend their leisure time doing anything remotely like their professional activities. In my previous working life I don't recall any of my insurance colleagues ever saying that they were nipping out to the travel agents to book an insurance holiday.

To support all of this expertise, there are vast libraries of books and research papers, as well as terabyte upon terabyte of information available on the internet. If you want to find out about a technical aspect of the role, then typically you will be spoiled for choice, advice and opinions. Of course there are still some technical subjects that not too much is known about, and hence little exists in terms of literature. For many this is all part of the fascination and challenge of working in ecology. So, with countless people to ask, and plentiful guidance about species, their identification, ecology and biology, today's ecologist has more to refer to than ever before. However, for many, the skills that are required in relation to their office working environment are often not given anywhere near as much attention. Sadly, most further education bodies do not cover many (if any) of the core business-related skills that an aspiring ecologist would find beneficial – or indeed, at times, crucial. It is often the case that the graduates who have more confidence and the appropriate skills to start a career within the business world are those who gained these skills as a result of jobs they carried out while trying to fund their way through university, as opposed to anything they learned during term time.

Chapter 1
BEING EFFECTIVE

Effectiveness is doing the right things.
Peter Drucker (1909–2005)
American management consultant, educator and author

I suppose the best place to get started is to define precisely what I mean by an 'effective ecologist'. Naturally, someone working within the ecology sector needs to be knowledgeable, competent and experienced in the taxonomic groups, habitats and survey methods they are involved with. In other words, they would be expected to have the relevant technical knowledge (including qualifications) and work-related experience.

Technical knowledge

Looking at technical knowledge first – well, that's a given, isn't it? Surely if a person doesn't have the knowledge then they shouldn't be in the job in the first place. Alternatively, they could be given the training required. But if neither the knowledge nor the training is in place, the employer should not be asking them to do something which is beyond their technical skillset. For example, someone who doesn't know how to identify plant species shouldn't be anywhere near a National Vegetation Classification (NVC) survey.

Work-related experience

Let us now consider the difference between knowledge and experience. It's not difficult in our sector to see that someone can have quite a lot of theoretical knowledge about a particular subject, without necessarily having any relevant practical experience. Many years ago I was quite well informed about cetaceans, and I could have told you huge amounts about whales and dolphins: where to find them, how to identify them, diet, breeding behaviour. Was I suitably experienced to carry out cetacean surveys? No, I most certainly wasn't. I had read lots of books and watched lots of videos (there's a clue as to how long ago I am talking about), but I hadn't seen that many cetacean species in the flesh in their natural environment.

It is clear, then, that knowledge and experience are not necessarily the same thing. A combination of both is essential in order for an ecologist to be capable of carrying out their responsibilities at the required level. So there you have it, technical knowledge and relevant field experience – that's all it takes, isn't it? If only it were that simple.

Effectiveness

Although this book is not going to make frequent reference to the technical knowledge and experience aspects of being an ecologist, it is very important for me to stress something at the outset. Everything that we will go on to discuss assumes that a person employed or engaged to carry out a professional job is knowledgeable, skilled, technically competent, qualified, experienced – and whatever else you want to add – at carrying out the role. In fact, when interviewing a candidate for a job in our sector these are usually the areas that are given the most attention. You can see and touch a qualification; you can discuss experience at length; you can talk about the type of sites worked upon and knowledge of different species.

It is much harder, however, for a potential employer to gauge the candidate's generic business behaviours, their interpersonal skills, and whether or not they are going to be a good 'cultural fit' for the business. This latter aspect is high on the agenda for any smart employer. As Steve Jackson-Matthews (Land Use Consultants Ltd) says, one of the most important criteria he assesses potential employees on is, *'Will they be a good fit into my existing team?'* If, as an employer, you have someone in front of you who is technically brilliant but your gut instinct is telling you that they won't be a good fit, then you are taking a big risk employing them. It is considerably easier to train someone lacking in technical ability than to squeeze the square peg of an unsuitable new staff member into the round hole in an effectively operating team environment.

Even when everything on the face of it goes well at the interview stage, it is not easy to be sure how well the interviewee will integrate into the team and whether or not they can be *effective* in carrying out the role. Are they really the workaholic they profess to be? Will they really pull out all the stops to meet a critical deadline? Are they really a good team player? What is pretty certain, though, is that as an employer you don't know what you and your team are really going to get until this new person walks through the door at 9 o'clock on their first morning. Assuming, that is, that they are on time in the first place. As the weeks and months pass by, will the new recruit actually prove to be effective in their role? So perhaps, after all, there is so much more that matters, over and above technical knowledge and work-related experience.

Of the three areas touched upon (technical knowledge, experience and effectiveness), I would suggest that the foundation to everything is effectiveness (Figure 1.1). Perhaps at this early stage you wouldn't entirely agree. So just in case you're not convinced, let's explore the matter a little further.

From your employer's and/or a customer's perspective, if you are the best ornithologist in your area, with a huge amount of knowledge and experience, but

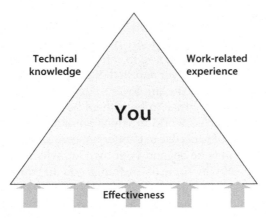

Figure 1.1 Effectiveness is the foundation of everything else

you communicate poorly and you never arrive on time, would that not be a concern? Let's add that you often forget to take the appropriate health and safety PPE (personal protective equipment) onto site and your reports are not only poorly written and full of errors, but also habitually late (in fact, very late). If this is you, the chances are your reputation will be severely damaged and you won't have a job or customers for much longer. In fact, if this is your approach to your work, who could actually say that you are the best ornithologist in your area? The requirement for any job would probably state that it needs to be completed safely, on time, within budget, and with a good-quality 'fit for purpose' report. If you are not achieving these things, are you really the best person for the job? Are you really the best ornithologist available to that particular employer or customer?

PERSPECTIVES

It is important, when engaging with other people (as we are all doing, all of the time), to bear in mind that the world around you often looks different from the perspective of others. As soon as you have more than one person's point of view you have the potential to have a difference of opinion, even over the smallest of matters. In the broadest sense, the way that you may be feeling internally on being told something could be very different to how the person giving you the message is intending to make you feel or perceiving your reaction. Such scenarios often hinge on how a particular event impacts upon you, bearing in mind your own perspective of how you will be affected, or not, by what is being proposed. For example, if you were told by your manager that your company vehicle was being sold and you now had to share a pool car, I imagine that you would feel quite negative about that. On the other hand, the situation from your boss's point of view might be very different. Money will now be available for a positive impact elsewhere in the business. What if the money saved could now go towards employing an assistant ecologist directly supporting you and removing some of your workload? How do you feel now?

Two sides to every story

As you continue reading this book there are two important, different and sometimes conflicting perspectives that I am going to refer to while exploring the topics and examples given: that of the employer and that of the employee.

First of all, let's look at it from the employer's perspective. How do you coach and give guidance to someone in order to help them be a better professional and an effective fit within the business world in which you are operating? It's difficult. It involves numerous areas of expertise (e.g. interpersonal skills, organisational skills, time management) that may lie beyond your own knowledge and experience. You may not even be aware of why some of the things you do yourself impact upon the team around you in the way, either positive or negative, that they do.

Secondly, let's consider an employee's perspective. What areas can you work upon and what things can you do in order to be more effective in your role? If you were more effective, would that have a positive impact? Would your life be less challenging, less stressful, more productive – and therefore more enjoyable, more satisfying and ultimately more successful?

The unfortunate fact of the matter is that most of us have been educated or have acquired our skills very much with the technical aspects (such as species identification and field skills) being the main or even the sole focus of our development. How many ecologists (at any level) do you know who have ever been given specific training relating to interacting with people? What about subjects such as management skills, effective leadership, communication skills, time management, project planning, negotiation skills, sales skills or finance? The number is very small in comparison to the number of people out there today who are being expected to perform many of these skills to a high level as part of their role.

Why is this the case? Well, there could be many reasons, but by and large it is usually the following. First of all, your employer may not know that much about these things themselves, and whether they do or they don't, they may not realise that you and the company could benefit greatly from better guidance. Secondly, a lack of available funds. And finally, pressure on time and resources. For example, you may have a training budget available, but who is going to want to go on a time management course when they can learn how to identify grasses and sedges? Perhaps if they were better at time management they would be able to do both.

Whether you are an employer, a manager or a team member in the ecology sector, the chances are that any non-technical skills you have been fortunate enough to pick up along the way have been acquired 'on the job' or as part of some other aspect of your life. It is less likely that you have been given any sort of formal guidance. This book helps to plug that gap. It doesn't cover everything, but it's a good start. The approach that I take is by no means the only way to do things. But if you take on board what is discussed in this book you will be more effective in your role, at whatever level that is. Note once again the key word: *effective*.

WHAT IS 'BEING EFFECTIVE'?

Good question, and thanks for taking me back to where I was a few minutes ago. Let me give you something to ponder over. Ecological consultants are very much in the service sector, but many of us fail to realise this. We see ourselves purely as ecologists, and forget (or, in some instances, may never have considered in the first place) that we are providing a service, managing risk and producing solutions. However, what we do is no different to what is offered by almost every other service-sector business out there. We have customers who pay for our services, and who rely on us delivering an excellent product.

What can these customers do if they don't receive what they believe to be an appropriate level of service? Simple. As has always been the case, they can go elsewhere. And in this age of the instantaneous alternative (easily found via the technology at our fingertips) it is easier than it has ever been to do so. So we are all operating within the service sector, and it just so happens that the service we are providing relates to ecology. And if it happens that you never directly see or speak to a customer, don't be fooled into thinking that you are not part of the service. Your role might be purely bat echolocation analysis or sitting on a remote mountainside carrying out vantage-point bird surveys, but if you fail to deliver there is the potential for a poor customer experience.

An effective employee

How would I respond when someone asks me what an effective ecologist looks like? Although not exhaustive, the following list should give you a good feel for what I mean.

An effective ecologist is someone who ...

- behaves professionally and fully understands how to operate with positive effect within the business in which they work.
- communicates appropriately within their team and with customers and suppliers.
- works within the prescribed systems, processes, methods and agreed budgets.
- has the vision to anticipate when a problem is looming and takes meaningful action to prevent the issue arising.
- listens well and understands precisely what they are being asked to deliver.
- completes their tasks on time and at the required level of quality.

An effective employer

Now let's develop the concept of effectiveness one step further, by considering it in relation to an employer or manager. What does an effective employer within the ecological sector look like?

First of all, assuming that they themselves are carrying out an ecological function at some level, then all of the aforementioned would still apply. What else? OK, you would be hoping to see good managerial skills and leadership behaviours. In addition to these, however, there is a huge part of the manager's role which involves delivering the required results on behalf of the business (Figure 1.2). Something that is often not appreciated, when considering that how these results are achieved, relates to the people working within the team. If the manager doesn't recruit the right people and develop the team members to be effective, then this certainly will impact upon what is achievable. Therefore, creating, developing and retaining teams that are going to be effective at delivering the required outputs is essential. Some people at this level in our sector do not think about it in this way. They may only be involved in team development (e.g. recruitment) a couple of times a year, and so they don't necessarily consider the positive everyday impact they can have on employee development and retention.

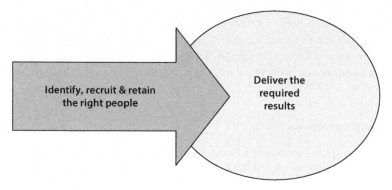

Figure 1.2 The right people will get the right results

My conclusion is that an effective ecologist in a managerial role is someone who is good at identifying, developing and retaining the right people. It is therefore vitally important that the person in that position is effective at recruitment, staff development and team building. If they get these areas right then the business will benefit greatly. Life will be so much easier, more productive and more rewarding for everyone involved, as the whole team moves forward in unison, rising to successfully serve customers and resolve daily challenges.

Robust foundations

All of that seems pretty straightforward, doesn't it? Well, if it is that straightforward why is it that so many people fall short so many times, and even the best of the best fall short some of the time? It is all very well believing that you can do all of these things well, but actually being able to deliver, time after time, day after day, against all of the pressures (both business and personal) that you face, is by no means easy. Therefore, rather than it being a haphazard *'some days I'm good, some days I'm not'* lottery, you should at least consider that there might be approaches and techniques that can be adopted and put into daily practice. As you develop the habit of employing these techniques, they all become part of your 'subconscious competence' (Table 1.1).

When you get these foundations to your business day fixed firmly in place you will find that so much of what used to be potentially contentious or cumbersome begins to evaporate. Ultimately, it's all about being effective in your working environment, for the benefit of your employer, their customers, their suppliers, your work colleagues, and of course yourself.

Table 1.1 The four stages of competence (adapted from Gordon Training International, undated)

Stage 1 Subconscious incompetence	Stage 2 Conscious incompetence	Stage 3 Conscious competence	Stage 4 Subconscious competence
You don't know that you don't know. There is lots more to a subject than you are yet able to perceive.	You are now aware of what you have still to learn, and you are making a conscious effort to develop your knowledge or skills accordingly.	You have now developed the skills or knowledge and you are in the early days of beginning to apply this, but you still need to think about what you are doing quite carefully while undertaking the task.	It is now part of your everyday knowledge or skillset and you aren't really thinking about it too much. It is just happening as part of your natural behaviours.

This model relates to how people progress towards being fully skilled, competent and/or confident at a particular ability, task or process.

Customer satisfaction

With all of this in mind let's look at Case study 1.1. The example given does not fit into an ecology box but nonetheless, using a fairly straightforward everyday customer experience, it gives us all something we can relate to.

Case study 1.1   It's all part of the service

My mum just had her garden landscaped. It took a couple of days, four landscape gardeners and a fair amount of money. It's not that she has a large garden, it's just that it was a bit of a mess!

After it was all finished, I was talking with her on the phone and I enquired about the progress that had been made. Her response was that the gardeners were fabulous, she would have no hesitation in using them again, and she was going to recommend them to a couple of her friends who were also needing work done. I then asked, 'Tell me what you mean by fabulous?' Here is her answer:

They arrived on time each day, they took their shoes off before coming into the house to discuss things, they took all of the garden rubbish away with them and they worked constantly during the job (no random breaks or disappearing off to do something else). There was one small problem, though. I asked them to make sure that when they had finished using the outside tap, to slacken it slightly, as they had turned it off too tight for me to use myself. They assured me that they would do this, but unfortunately they didn't. I noticed this the morning after they had finished the job. I was going to have to ask a neighbour to sort it for me. But at lunchtime that same day one of the gardeners appeared at my door to say that she had remembered late last night that they hadn't slackened the tap. She apologised and was embarrassed that they had forgotten to do this, and she was now here to fix it. She went into the garden, slackened the tap appropriately and left, apologising once again.

All that way just to turn a tap! But it wasn't a tap that was the issue here, it was a 'promise'; it was a reflection on their integrity and their professionalism; it was all part of their service ethos; it was all to do with customer satisfaction.

Now the thing to notice in this real-life example is that nothing that my mum said, regarding her satisfaction that the job had been done well, related to how knowledgeable the workforce were about the plants they were planting, the trees they were cutting, the slabs they were laying, and so on. All of that was taken as a given. What differentiated those workers all related to how they interacted with their customer, how they were organised (arriving on time and working within budget), and how they followed up, and resolved, a potential issue before it became a real problem. They got the basics so right, at such a high level, that overall they were totally effective.

EFFECTIVE VERSUS EFFICIENT

We have now spent a bit of time looking at what *effective* is all about, from an employee's and an employer's perspective. However, how does *effectiveness* relate to *efficiency*?

The first point I would like to make is that someone who is effective would normally have an efficient approach to their workload. However, achieving efficiency alone does not always equate to being effective. I think we have all experienced heavy workloads on our desks and a customer promising to appoint us tomorrow, for a survey that should have been carried out a month ago, to produce a report that was needed yesterday. Ecology isn't easy. We are one of the few service-sector businesses where a huge part of what we deliver is governed by the seasons. If something gets missed (for example, a great crested newt survey) it may be many months, or even longer, before it can be revisited and a solution provided. Everyone else involved in the process (landscape architects, project planners, planning consultants, construction companies and the like) can usually start work and make progress at almost any time of year. All of these other professionals must find the ecological aspects of their projects quite frustrating at times. So efficiency is definitely going to be a major factor in determining how effective we ecologists are for our customers. Can we get the survey done at the right time of year, and thereafter can we deliver a 'fit for purpose' report to the client by their deadline?

What is efficient? When most of us hear that word we think 'speed of process'. Often, in order to achieve speed (i.e. efficiency) in a service-sector scenario, there is a sacrifice, and that sacrifice usually relates to quality. So when we are really up against it, what ultimately do we sacrifice: speed or quality? The quicker the job is done, the greater the risks are that something somewhere hasn't been properly covered or fully considered. You therefore have this potential for conflict. *'Yes, I can do it faster, but it won't be as good.'* Sometimes there may be a way around this. It might, for example, be possible for the customer to pay more in order to achieve both speed and quality (e.g. double the resources working on the project over the shorter period of time). Let's be realistic, though: even on the rare occasion when a customer is willing and able to pay more, where does an ecological consultant instantly find reliable, effective expertise at the drop of a hat, mid-season? Anyone out there who might be able to assist is probably up to their eyes in their own seasonally constrained challenges. So let's take the *'I can pay more'* scenario out of the equation.

Being fast, and therefore efficient, does not always equate to being effective, although undoubtedly someone who can achieve both, when under pressure, is gold dust. So please do not mix up the two. There is definitely a conflict, and if you are not on top of it, it could very well be your undoing. Usually in our world the biggest threat to quality of output presents itself when we are running out of time. Being effective, in this case, is about understanding, early on, the situation that is looming. Then having identified that there is a potential problem it's a matter of making the right decisions in order to help reduce the risks associated with running out of time or missing the deadline. Having made these decisions you must then communicate with everyone else involved to ensure that they all know what's happening and why. If the deadline can't be met, then you must ensure that the people who are relying on you are made aware of this now (not later today, or tomorrow, or an hour before it's due – lift the phone now). Yes, ultimately a deadline (assuming it is a critical deadline that cannot be shifted) is a deadline, but you will be surprised how often a deadline set some months ago gets shifted back and no one thought to tell you (how ineffective is that?). Ask the question. Is there some wriggle room, can we renegotiate, can we stagger the outputs – for instance, if we can give you the bit you really need tomorrow, can you wait another couple of days for the rest?

Speed or quality?

You may have noticed that so far I am precariously perched on a barbed wire fence as to what ultimately gets sacrificed – speed or quality – when there is no other way. There isn't really a right or wrong answer. There are so many different scenarios that would draw me in opposite directions. All through the thought process, however, I would be asking myself, *'How did this happen, and how can we prevent a repeat occurrence?'*

Often the answer is not so much about today's challenge (short-term thinking), but more about the future development of the people involved. Additionally, systems and processes may need to be investigated, as they could have fallen down and need to be fixed, or new procedures may need to be put in place in order to prevent reoccurrences in future projects.

At all costs you must remain credible in your dealings with your customer and give them the best that you can possibly deliver within the time constraints. Some ideas will be forthcoming elsewhere in this book as to how you can manage scenarios such as this. Ultimately you are working in a sector where you are often required to hit tight deadlines while still producing quality outputs. Usually, when it goes wrong, it's because somewhere in the process (perhaps even because of how you have managed your own time in the weeks leading up to this point) someone has failed to notice early enough that something isn't going to be achievable. Then by the time it is communicated it is too late to find reasonable alternative solutions.

The answer to dealing with the challenges you are often faced with is to anticipate when problems may arise, far enough in advance in order to give yourself or your manager the opportunity to fix it before it's too late. Simply put, it's down to paying attention to what you are responsible for and communicating effectively (this latter aspect is covered in Chapter 4). In your working environment, are there ways in which this becomes more easily achievable? Of course there are. You just need to build a workflow system that monitors and reports upon progress throughout the lifespan of a project. Someone (a manager) needs to be responsible for putting the system in place and ensuring it is adhered to. Those working on cases within the business then need to be operating within the system and communicating with everyone else involved, every step of the way. If the system fails at any given time then a review of what went wrong needs to quickly take place, and adapting the process in order to prevent the same problem happening again will be required. Chapter 7 (*Project management*) will give you some ideas in this respect.

TAKING STOCK

In conclusion, being effective, whatever your level in an organisation, has an immense impact upon whether or not you are likely to be successful within the various roles you are going to take on during your career. Whether these roles are regarded as having been successful, or not, is determined by the results you achieve on the projects and challenges that are placed in front of you. In understanding what effectiveness looks and feels like, it is important not only to consider your individual perspective as an ecologist, but also to fully take on board the perspective of your employer and manager. Conversely, if you are an employer or manager, you should be considering the perspective of your employees. Customer perceptions are also very important. Ultimately it is your customers who will use the work you have produced, and they will constantly be considering how effective you actually are. Being effective impacts upon service levels, customer retention, success in business and careers.

As we are about to delve deeper into the various subjects covered in this book, let's ask an important question. Do you want to be better, do better and achieve more? Do you want to be successful? In order to help achieve these goals, being more effective can only have a positive impact. We can all aspire to improve, and I hope that in the remainder of this book I can put across ideas that help us move forward professionally to the benefit of ourselves and all around us.

Chapter 2
WHAT'S THE JOB?

Don't try to be different. Just be good.
Being good is different enough.
Arthur Freed (1894–1973)
American lyricist and Hollywood film producer

Before I answer the question *'What's the job?'*, I would like to discuss the events leading up to someone getting a position within an ecological consultancy in the first place.

THE PREQUEL

In the ecology sector, getting a paid position can be tough. Having completed a degree or obtained some other suitable qualification, many people struggle to immediately find a full-time permanent position. In most cases they are faced with *'You don't have enough consultancy-related experience'* – and if I have heard the frustrated *'I can't get a job without relevant experience, but I can't get the experience without a job!'* once, I have heard it a hundred times. For many this means that they set off along the path of post-university education, seeking self-development in relevant subjects. In doing so they may attend training courses, join clubs or societies, or volunteer. Eventually they get seasonal work, then perhaps they move on to a longer fixed-term role and finally, hopefully, a permanent position. Along the way a fair number give up and divert course onto other career paths.

When seeking employment for the first time within the sector, a graduate very often does not fully appreciate what working within an ecological consultancy entails. It is not very likely that they will have covered this in much depth, if at all, within their studies and it would therefore make lots of sense for them to thoroughly investigate what the role of a consultant is. There are a number of common misconceptions that may occur at this stage. To quote Steve Jackson-Matthews:

> An ecological consultancy is not there to prevent development occurring, and conversely it is not there in order to support every developer with everything they want to do. The position is far better described as being one

of relative impartiality, whereby the consultant 'consults' with all parties involved, at every stage of the process, in order to arrive at an acceptable outcome. All of this needs to bear in mind the ecological credentials of each and every case.

Another point that should be seriously considered by a potential new recruit is that a career within our sector is unlikely to be a nine-to-five, 35-hours-a-week affair. Once you have found a job, the hard work is only just beginning. You may think that you have been working hard up until now, but usually new entrants find that what is expected of them in a typical working week (especially during the busy spring/summer seasons) is well beyond what they may have envisaged. You need to go into this career with a fair degree of flexibility regarding your availability to carry out work for your employer. It would be very unusual if you were not required to work late nights or early mornings. If you think about when the sun sets and rises, and then factor in travel time as well as complying with the various survey methods, that will begin to give you an idea of a typical working day (or night). Steve described it well to me when he said:

> They have their whole career in front of them. It's these early years, doing the hard graft in the field, when people learn the most. This gives them the experience and understanding they will need in the future when they themselves are in a more senior position and need to be effective in their decision making. This is partly why it's not realistic for new entrants to expect to progress quickly up the ladder to a senior ecologist position. A senior ecologist has a huge responsibility, and their employer has to be confident that they fully understand the implications of the numerous important decisions they may make in a typical week. People coming into the sector for the first time should just take a deep breath, work hard and focus on enjoying the experiences that these early years will give them. This will equip them well with what they will most definitely need later on in their working life.

Numerous books and resources exist that cater for people seeking employment. For example, writing your CV and performing well at the interview stage have been covered elsewhere so thoroughly, so many times and in so many ways that I am not going to attempt to add much to these areas here. One thing I would say in this respect is that very little published literature appears to exist specifically aimed at the ecological sector. One exception that I have come across in the UK is an excellent book by Susan Searle called *How to Become an Ecological Consultant* (Searle 2011). For anyone seeking employment within our sector I would thoroughly recommend reading that book.

The recruitment process

As an employer, quite often when we are faced with graduates applying for their first position, we are presented with applicants who are not work-ready. That is, not unless they have taken it upon themselves to develop skills and gain experience beyond that provided by their university tuition. This means that we have to ensure that the person we take on has the skills to add value to the business from the start. Hence, previous experience is eagerly sought. Alternatively, we can employ someone who will need quite a bit more attention and development before they are in a position where they are able to add value in their own right.

Every time an employer advertises that they are recruiting, at whatever level, they have a reason for doing so and a strong idea as to what their perfect applicant should be capable of. Once an advertisement goes public it's not long before the recruiter has a lengthy list of candidates responding with their CVs. It is now time for the next stage. Who will be invited for an interview?

What is actually going on at a job interview is usually quite simple. The employer is looking for the best person they can find to solve a problem (or potential problem). Be clear on this. The employer isn't doing this with the motivational emphasis of furthering the candidate's career. If there is no benefit to the business, then there is no point. It might be that the employer needs to be more efficient with workflow, or has a skills gap that must be filled. It could be that they are growing their business and need more people in order to deal with the greater workload, or they might have grown to a point where they need additional people at a managerial level. But remember, if there were nothing in it for the employer then they wouldn't be putting themselves through the distraction, effort and expense of looking for new people.

Of course there is also the perspective of the employee to consider at this important stage. When the right employee makes the right move at the right time into the right role, they get what they are after, as well as the employer getting the solution they require. Figure 2.1 gives a flavour of what each perspective may very well be at the outset, when a new position becomes available within our sector.

WHEN CAN YOU START?

What about the next part of the process, though? What should be done once someone has landed the job? Moving on from this 'Can you start on Monday?' moment, not that much published guidance exists for any sector, let alone our own. Let's be honest here, though – the lucky applicant is probably so relieved and excited at this point that even if there were material out there they wouldn't feel that they needed it. Of course they would be misguided in thinking like this, but let's not take the shine off the occasion just yet. Congratulations, well done! It's time to celebrate. They are fabulous, after all. They have just been through a tough interview process and beaten off all the competition. It has been decided by their future employer (arguably now the most important person in their life beyond close friends and family) that they were the winner.

Once the celebratory night out with their pals has passed and the start date draws closer, and once they have abseiled down from Cloud Nine, let's inject a dose of reality. Having been successful in acquiring a position, the excited new team member and their new manager sometimes forget that although it may be the end of one

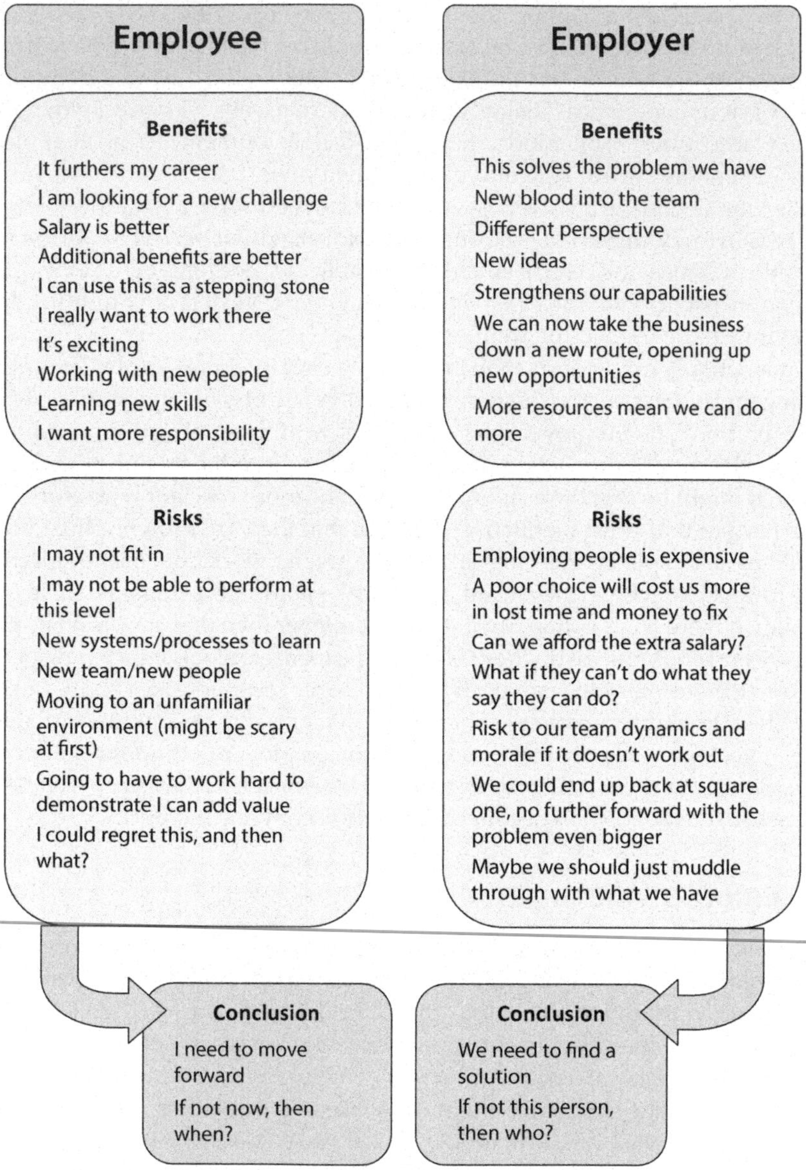

Employee

Benefits

It furthers my career
I am looking for a new challenge
Salary is better
Additional benefits are better
I can use this as a stepping stone
I really want to work there
It's exciting
Working with new people
Learning new skills
I want more responsibility

Risks

I may not fit in
I may not be able to perform at this level
New systems/processes to learn
New team/new people
Moving to an unfamiliar environment (might be scary at first)
Going to have to work hard to demonstrate I can add value
I could regret this, and then what?

Conclusion

I need to move forward

If not now, then when?

Employer

Benefits

This solves the problem we have
New blood into the team
Different perspective
New ideas
Strengthens our capabilities
We can now take the business down a new route, opening up new opportunities
More resources mean we can do more

Risks

Employing people is expensive
A poor choice will cost us more in lost time and money to fix
Can we afford the extra salary?
What if they can't do what they say they can do?
Risk to our team dynamics and morale if it doesn't work out
We could end up back at square one, no further forward with the problem even bigger
Maybe we should just muddle through with what we have

Conclusion

We need to find a solution

If not this person, then who?

Figure 2.1 Employee/employer recruitment perspectives

process, it is now very much the beginning of an even bigger one. Up until this point no one was really totally committed to the situation. The employer had the option to offer the position to someone else and, of course, the candidate had the option not to accept the offer. However, in a heartbeat that has all changed. An offer has been made and accepted. There is now commitment on both sides. So where do we go from here?

The interview hasn't really ended yet

The employee would do themselves no harm whatsoever by taking an approach which in effect says that *'the interview hasn't really ended yet'*. In fact, if anything, matters have just got considerably tougher. Everything now goes up a level and the stakes get higher. Now there is the potential for something to be taken away from them (their job). Also, from the perspective of those who recruited the new employee, their reputations are on the line. Their team, their peers and their bosses could all be impacted upon, or have something to say, if it is eventually shown that a poor choice was made. And it's not just a matter of reputations here. As I have already said, it's expensive employing people.

New team member – the employer's perspective

If you consider all of the expense leading up to the new recruit's first day (the cost of advertising, the time spent reviewing CVs, the interview itself, the fees payable to the recruitment consultant, and so on) you will find that costs of many thousands of pounds have already been incurred. Once the new person walks through the door many more thousands are about to be spent as time is set aside for induction, training and integration into the business. Furthermore, there is the cost of additional equipment that may need to be purchased, and of course their salary and pension arrangements.

The point I am reinforcing is that it is most definitely an expensive business, and in the world of business, a business owner shouldn't incur an expense unless they are going to be better off for it (e.g. financially beneficial, more secure, or more efficient). Therefore, the cost of getting the decision wrong can be substantial. In addition, there are emotional costs and distractions for the manager concerned, ranging anywhere on the scale from inconvenient through to extremely stressful. Then what happens if the person doesn't begin to make progress towards the desired level and either leaves or isn't kept on? The whole recruitment rollercoaster starts again.

New job – the employee's perspective

As an employee, you should consider that you are always being assessed at some level by someone. It may be your manager watching your time keeping, or someone reflecting on your reporting ability. You are in a commercial world, and somewhere in the organisation someone is looking at the performance of the business overall (sales figures, expenses, profit, cashflow, staff retention, etc.). The business comprises the component departments, the teams, and the individuals who make up those teams. Monitoring will therefore, more than likely, be taking place at department, team and individual level. When things are going well, somewhere someone will be assessing why, and if things are dipping in a particular area, then again someone will, at some stage, be asking the question: why?

As a new employee, you need to consider that you are now getting paid in order to perform at a certain level. You are in a business environment and that demands (yes, demands) that you perform successfully.

SO WHAT IS THE JOB THEN?

'What's the job?' is an expression that comes up quite often in an effective business environment. In the context of day-to-day business activities it's often said in order to reaffirm at any point in time what precisely is required. For example, if an organisation is commissioned to undertake a bat survey of a structure, the job quite simply could be to identify any roosting locations for bats within the structure, species involved, numbers present and type of roost. It could be no more than that. With that in mind, documenting bats foraging and commuting in a nearby woodland would not be a required deliverable for the client, as much as it might be of interest to the ecologists surveying the site.

Sometimes, when everything around you is hectic, and with all the other challenges presenting themselves along the way, the rationale behind a specific job gets watered down or misted over. Occasionally the task instructions are not properly conveyed, and those undertaking the job do so ineffectively, or are misguided, or, worse still, are totally 'off the mark' from what is required. At times such as these it makes sense to just step back from it all, adopt a higher perspective and remind yourself, and everyone else involved, what precisely you are being asked to do and why. It is a bit like pressing a mental reset button on the whole project. Sometimes you just have to bring it all back to basics. OK, all of that seems fairly logical when considering the day-to-day responsibilities of an ecologist. But let's now develop matters further and consider the whole concept of *'what's the job'* at its highest level (in other words, thinking beyond the specific tasks that are on your desk). A useful way to think about this is in the context of an ecologist being seen as successful within their business environment.

Navigating your way through the interview process and ending up with a decision on both sides – to make an offer and to accept the offer – is only the first step forward in the process of achieving the desired result for all concerned. What ultimately is that desired result? Surely it is for the recruitment process to have found the right employee for the business, with this manifesting itself in terms of their performance. Adopting this approach leads us to the conclusion that, at the highest level, the answer to the question *'what's the job?'* is as follows:

> From the **employee's** perspective, the job is to keep your job and ensure as far as possible that you are successful in the role that you are being paid to perform. Also, it is to ensure, on the whole, that it is a positive experience that is giving you the right level of momentum, development opportunities and challenges. And all of this is in order to feed your passion for the career you have chosen for yourself.

From the **employer's** perspective, the job is to ensure that your new employee performs effectively in the role that has been given to them in order to help the business succeed in achieving its goals.

Everything else that happens in an ecologist's day-to-day business life (e.g. all the surveys, the reports and the analysis) sits beneath those two key definitions of 'the job'. The employee's definition won't be at the forefront of your mind while you are sitting there typing reports, constructing species protection plans or working out how you can find the time to carry out yet another bat survey next week. Likewise the definition as seen from the employer's perspective will not always be at the forefront of your manager's mind (thank goodness they have so much more to think about!). That is, not until something is not going well or not going to plan or is not being delivered as it should be.

In order to satisfy both perspectives, probably the most vital area where both the employee and the employer need to perform at the highest level is effective communication (see Chapter 4). The manager must make every effort to communicate clearly to the employee precisely what is required. Both would benefit greatly from an environment in which they can exchange thoughts and ideas about how the employee can improve their performance and develop within the business.

Taking on board all of these points, the answer to the question *'What's the job?'* is developed further in Table 2.1. You can see that the employee's objectives are not hugely different to those of the manager, and for each of them everything is in fact mutually beneficial.

Overall it's a fairly straightforward concept to get your head around, whether you are an employee or an employer. Far less straightforward is the answer to the question that may come from either side: *'So how do we do our best to ensure that all this happens?'* One would hope that a manager in our sector would already have the skills (and guidance if necessary) to help them fulfil what is required of them. Accordingly, in answering the question now posed, I am going to give guidance purely in respect of the employee.

Table 2.1 What's the job? Two perspectives on key objectives

Employee's perspective	Employer's/manager's perspective
Keep my job	Help ensure new employee is effective
Perform at the required level	Monitor performance
Communicate effectively	Communicate effectively
Be a valuable asset to the team	Integrate new blood into the team
Support my manager	Coach, encourage and give feedback
Develop my skills	Create opportunities for development
Progress to the next level	Allow for progression when time is right

Making it happen (the employee)

There is no single magic formula, and different approaches may need to be deployed in different workplaces, at different times, in different scenarios, with different employees and with different managers (he or she that holds your fate in their hands). However, there are lots of things you can consider, apply and develop within your working environment to help you perform at your absolute best within the role you have been given. Having said all of that, of course, there may very well be an occasion when you lose your job because of something that was totally outside your control. Something beyond the walls of the business has conspired against you, and the employer has had to let you go, no matter how good you are. I once had to let a really good ornithologist go because the contract for which he was employed was terminated as a result of one of our customers going bust. He was unlucky. He was the right person with the right skills, in the wrong place at the wrong time.

Setting aside all of the 'uncontrollables', you can, however, take a positive approach by focusing wholly on the 'controllables' (see Chapter 3, *Positive behaviours*). These are the things that you can influence, the stuff that will make you a key team member, or, even better, indispensable to your employer. You can't do anything about customers going bust or your business owners adopting a different strategy; don't drown yourself in matters beyond your control and influence. Instead, immerse yourself in controlling the controllable; work hard and work smart. Focus on being good at what you do and on being effective at how you deliver success for the business.

At a very basic level, if you want to keep your job then get yourself into a position such that if your manager ever needs to get rid of someone, they will think long and hard about how much more difficult their life would be if you weren't around. As I have said on numerous occasions to new starts on a fixed-term contract within our business, '*Make me regret the day after the last day of your contract. Make me and our team miss you when you are gone, and you know, you just might get lucky. You might just be the right person, at the right time, in the right place.*' I have just heard a distant cynic cry, '*Yeah, that will be right.*' Well, as I type these words, our business has two people working with us today who have just had their fixed-term seasonal contracts extended to annual contracts for that very reason.

Make your manager's life as easy as possible

It is also good to adopt the opinion that your job is to deliver what your manager requires of you, and that you should be endeavouring to make your manager's life as easy as possible. That includes making sure you do everything in your power to make your manager look good in the eyes of their boss. Those who really excel in this effective behaviour regularly anticipate and react to what is required before their manager has even considered that the task needs attending to in the first place. This could also, in some circumstances, be described as '*doing today what you anticipate will be required of you or someone else tomorrow*'.

Put another way, you definitely do not want to be doing things that make your manager or your employer look bad. Also, it most certainly is not your job to be making your manager's life difficult. It is not your job to be constantly raising issues or causing problems that your manager needs to attend to. Doing so could mean that your manager might not be able to devote enough of their time to all of those important tasks that generate a positive momentum and contribute towards their part of the business being successful.

Bearing in mind what we have just discussed, there are a couple of points I would like to make. Firstly, you must always steer well away from anything unethical or illegal, or anything that contradicts or conflicts with the overall goals of the business. Secondly, it must be acknowledged that there are some bosses who quite frankly have achieved their position despite themselves, and are not the kind of people that you would want to make look good. It may even be politically in your interest to not provide too much assistance to such a person, in the hope that they are eventually moved on. My advice, in this respect, is this: keep your head down and get on with your job without creating any conflict. It may be that in every other respect you enjoy working where you are, in which case you certainly don't want to put your job at risk. If, on the other hand, things are really bad, then perhaps, quietly, in your own time, the answer is to make the necessary moves to find work elsewhere. If you do move, then ensure it is on your own terms and when the timing suits you. Oh, the joy of office politics and human nature!

HELP! IT'S ALL GOING HORRIBLY WRONG

An area that I really must touch upon concerns what happens when things aren't working out for the employee. Naturally I want to stay as positive as I can throughout this book. However, unfortunately from time to time things do go wrong, and the least I can do is offer some advice as to what to do when faced with that situation. You might be wondering if this is the right time and place to be talking about this particular subject. I feel it is. A huge part of 'What's the job?' is actually ensuring that you keep the job, and that you are also well positioned to move forward with your career when the time is right.

There is nothing that will hinder your career progress more than being stopped firmly in your tracks and being told something like 'You are just not performing to the required standard.' Often at this stage things can then begin to go horribly wrong from both the employer's and the employee's perspective. 'Well, Neil, who cares about the employer? It's me that's going to be looking for a new job.' From your point of view you are of course right, but as always there are two perspectives at play here, and both sides of the table would be greatly advised, right from the start, to appreciate this. So let's proceed on this basis.

The employee's perspective

I am sure you won't be interested in hearing the employer's side just yet. You are the one that is feeling vulnerable, and it's really important that your feelings are considered and your opinions are listened to. Quite often it all starts with early signs that things are not entirely what they should be. Sometimes the employee is aware that they are struggling with some aspect of their role well ahead of the manager discovering it for themselves. It strikes me as odd that many people either fail to fully understand what a high standard of work output is, or know that they are struggling but fail to ask for guidance or coaching in order to improve matters early on. I suppose it's a bit of *'Don't admit I have a weakness'*, and a bit of *'They won't notice if I don't mention it.'*

Tackling the first bit, I would say that it is a strength to understand where your weaknesses lie and to be honest enough with yourself (and probably with others) that you need to learn more about some aspect of the job, or work harder in certain respects in order to improve.

Now the second bit. Do you really think that it won't be noticed? If you are working within a professional team, with a manager who knows what good performance looks like, you would be an idiot to believe that you are smart enough to consistently hide your poor performance. It may be that your manager just quietly accepts that your positives outweigh your negatives and chooses not to raise the matter with you. But please do not ever think that just because it's not being talked about it hasn't been noticed.

And the best solution? Taking it upon yourself to improve your own performance is a controllable – fix it. Either that or run the risk that one day it will become a huge issue, when it has happened too often for it to be redeemable. In short, take control and accept responsibility for your own actions or lack of actions, and be genuine and credible in any discussions you are having with anyone about areas where you require advice or coaching.

Now, quite frankly, some employers and managers go from week to week, month to month, and, horrifically, even year to year without carrying out any sort of one-to-one meetings (121s), performance reviews or feedback sessions (positive as well as corrective) with their team. Some managers find it very hard to accept that, almost without exception, it is better to tackle poor performance and the like early on and give corrective feedback when required. Others find it hard or uncomfortable to give praise when it's due. In the context of these two extremes – 'corrective feedback' and 'praise' – Reuben Singleton (Tweed Ecology Ltd) kindly made me aware of a phrase which accurately describes how a manager should deal with these situations. It was first used by the American football coach Vince Lombardi (1913–1970), who quite simply said *'Praise in public, criticise in private.'* There is so much that lies behind that statement, and a manager who consistently adopts this approach is very credible indeed. As an employee, you should do the same.

The employer's perspective

When it comes to having effective working relationships with the individuals in your team, as a manager you need to set up some system of dialogue with each team member whereby opportunities exist for you to provide guidance, review performance and deliver feedback. There is a lot of material and advice available about these areas, and if this was a book aimed specifically at managers I would have so much more to say about it – but here are some points that really shouldn't be missed.

In setting up your staff dialogue systems you should ensure that you hold a meeting with every team member regularly (at least monthly, preferably more often), and that you do so with positive intent. If you are one of these managers who constantly feels that a member of your team is letting themselves and your business down, but you don't feel able to tackle it head on, then please note what follows. If someone doesn't know that they are doing something wrong, they will not have the opportunity to fix it and they will not be able to change their behaviour. If your approach is to wait it out until it either fixes itself (it won't, unless it's raised) or becomes a massive issue, then ultimately you are mismanaging that person, because one day they may be required to go through a disciplinary process. You might be thinking you are doing them a favour by not mentioning the issue. Trust me, you are not helping them. In fact, through trying to be the nice likeable person (as opposed to an effective manager) you are potentially causing them a huge amount of harm later on, either with yourself or with a different manager elsewhere.

When most employees take time and really think about the risks and benefits associated with feedback, most good employees (genuine, hardworking, professional types) want to know where they can improve. The difficulty lies in that the whole concept of feedback (good or bad) hasn't really been discussed properly with the team member, or developed properly with the manager. Many managers have not been well trained in how to deliver feedback. And how many employees have been shown how to receive and react to it? In Appendix 1 (*Feedback: get rich quick*) I have provided some guidance on giving and receiving feedback.

Managing poor performance is all about taking action as soon as possible in order to try and resolve issues before matters get out of control. If things can be tackled appropriately, professionally and effectively early on, then the desired result is often still achievable. When this isn't the case and matters have to be taken further, for whatever reason, by either party, then unfortunately we move into the realms of performance improvement plans, disciplinary action, and resignations. These areas are beyond the scope of this book.

CONCLUDING THOUGHTS

In summary, your job as an employee is to keep your job and perform at the highest standard you can. As the employer or manager, you have the additional responsibility to ensure, as far as possible, that the employee is effective in their role and delivers what is required of them for the benefit of the business.

Whether you are an employee or an employer, at all times remember what you need to deliver. If your radar is telling you that things could be going wrong on a distant horizon, then take action to rectify the situation while it is still on the horizon and you can steer a carefully considered course away from trouble.

Chapter 3
POSITIVE BEHAVIOURS

The harder I practice, the luckier I get!

Jerry Barber (1916–1994)
American professional golfer

I have already touched on the benefit of focusing on the controllable, as opposed to drowning yourself in matters beyond your control (Chapter 2, *Making it happen (the employee)*). What are these 'controllables' that you can immerse yourself in? Broadly speaking they are skills and behaviours you can choose to adopt (or risk to ignore!) in order to help you on your journey and thus support you in performing effectively at the level required by your employer.

In this chapter I will examine many of these positive and controllable behaviours under three broad headings:

- Be professional
- Early days – establishing yourself in the role
- Established in the role

In discussing these themes I am going to avoid delving too deeply into areas covered more thoroughly elsewhere within this book, for example communication skills (Chapter 4) and organisational skills (Chapter 5).

BE PROFESSIONAL

There is quite a lot of debate about what is meant by the word 'professional'. In some contexts it simply means that you have skills or knowledge that you use in a business environment (i.e. you are getting paid for the use of your skills and/or knowledge). In that sense, all paid workers can call themselves professionals relative to the job they are doing. Accordingly, we have builders who are professionals, housekeepers who are professionals, accountants, lawyers, gardeners … the list is as diverse as it is endless.

Let's now ask the same question in a slightly different way. What is doing a job 'professionally'? Put like this, it takes on a different meaning. This implies that not only are we doing a job that we are getting paid for, but we are doing it to a particular standard. A standard that would be deemed at least acceptable (i.e. 'professional') in the eyes of a customer. Figure 3.1 will give you some guidance as to what being professional might entail in an ecologist, as seen by a customer.

When you consider further the content of Figure 3.1, you may feel that there isn't that much there that refers specifically to the ecology sector. Many of the qualities described could easily fit into other service-sector businesses. I have deliberately described it like this to highlight the fact that whether or not you are deemed to be doing a job professionally is very much in the eyes of your customer, and that, almost without exception, the ultimate customer for every job in our sector is not an ecologist and therefore would not normally judge your professionalism against a subject that they know little about (i.e. your technical knowledge). Or, if they do apply such a judgement, it would sit amongst a lot of other non-technical measures.

When I have a new client approaching our own business and they make comments (positive or negative) about other ecological consultants they have used, it is rare that the initial points mentioned relate to technical ability. Your customers aren't experts on ecology, and as such when thinking about you and your products (i.e. the services you provide) they will relate it to something that they do know about. And in today's world everyone knows or has something to say about customer service. Accordingly, it is really important that you fully appreciate how your customers will determine how well you are performing for them.

Hold on a minute! Most of us don't have direct dialogue or contact with the customer. Usually that is our manager's role, or a job for the senior ecologist, or we

Figure 3.1 The complete professional: a customer's perspective on ecological consultants

are acting as a subcontractor for someone else who is communicating with the customer. Well, the good news is that it doesn't really matter where you are positioned in the line. All of the qualities required in order to provide good customer service should still come into play at all times, no matter what. A good way to approach this is to treat the person you are reporting to as if they are the ultimate customer. For example, your manager is your customer. They have requested a service from you in return for payment (i.e. your salary), and behaving professionally will do your personal brand and credibility no harm whatsoever in their eyes. And when they are describing you to others (for example, to the business owner, when they are being asked if you are ready for a promotion), what do you think the bulk of that conversation might involve? Again, technical knowledge on its own just isn't enough to describe your abilities to others. In fact, if I asked the team manager in my own business about a member of her team and her response was purely, 'He is good at identifying mammal field signs', I could very well read between the lines and conclude that he had shortcomings in other respects.

Being professional means adopting and embracing the required qualities and evolving to be better today than you were yesterday. A good way to approach this is what is sometimes described as being 'in the zone'. You walk through the door at the start of a shift and you 'put on' your professional persona. This is because it isn't good enough just to know what you are doing. In the eyes of a customer, and your manager, you need to look like you are able to do the job, act like you know what you are doing, and sound like you know what you are talking about. All of these then support and feed into each other, portraying you in the best professional light (Figure 3.2). This all helps to build and reinforce the confidence that everyone around you has in your abilities. It allows others to feel comfortable that you will be able to get the job done to the required professional standard.

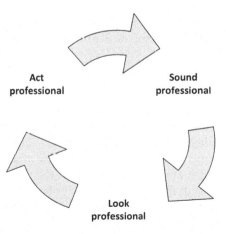

Act professional

Sound professional

Look professional

Figure 3.2 Be professional

When I spoke to Steve Jackson-Matthews about this, he developed the idea further by saying:

> Your integrity and the respect you have for the profession you are in and the work you are doing at any specific point in time are hugely important. This whole concept of respect extends right through to the customers you are working for, the company that employs you and, of course, yourself.

Part of the manner through which you demonstrate this integrity and respect is through your commitment to the role, the development of your technical abilities, and how you communicate with those you come across professionally.

Professionalism is therefore very much in the eyes of the receiver – in the messages you, your body language and your behaviours transmit while you are in your working environment. Professionalism in your day-to-day office environment may differ from what it is when you are on a building site doing an ecological clerk of works (ECoW) role, or from what it may be in a customer's premises when you are making a sales pitch to acquire a new project. Much of what is written throughout this book is about performing at your best and adopting those behaviours that are most likely to be deemed to be highly professional. Ultimately it is about doing your absolute best, to the required or better than the required standard every time, in order to deliver to your customer's expectations. In doing so you should always be conscious of how you may be perceived, and you should always ensure that you are attired suitably and communicating appropriately according to the circumstances that present themselves on any given day. The communication aspects of this will be covered more thoroughly in Chapter 4.

Truly professional

But is there even more to it than that? Until I really started thinking about it I would have said that the sorts of things shown in Figure 3.1 were pretty much what a true professional in our sector might look like. Then I was reminded by a colleague about something that had taken place some time prior to me writing this. Case study 3.1 describes what happened.

Case study 3.1   Just get on with it

I was delivering a training course in Birmingham. In the couple of days leading up to the course I had been quite unwell – and anyone who knows me will know that I don't do 'unwell' well. I had the dreaded, body-zapping, brain-burning, I-think-I-am-going-to-die (or at the very least, sneeze) man flu. But worse than all of that, my voice was going. Not a great place to be when you are about to stand in front of an audience for six hours. Anyway,

---

off to Birmingham I headed, a customer was expecting me to deliver, and no matter what, I would deliver.

The customer had arranged for a dozen people to attend the event, and this involved them booking hotel accommodation, purchasing rail tickets, etc. Now some of you may be thinking, *'Neil, you're an idiot. You should have just cancelled the event and arranged a new date. The customer would have understood.'* Trust me, things were so bad that this was seriously considered. But here is a different perspective.

If I had cancelled, then yes, the customer would have said that they understood, and that we could rearrange, and get well soon Neil – the operative word here being 'said'. Often what people say and what people really think are entirely different. If I had cancelled, then they would have had to spend hours unbooking hotels and trying to get refunds on items booked (including catering and training facilities). In addition my customer would have needed to let the delegates know that they would all have to do similar unproductive and inconvenient things: reorganising business diaries, social commitments, child care and goodness knows what else. All because the trainer had a cold! In amongst all of that unproductive time, and money lost, every single one of those involved would have had to be complete saints not to feel at least a teensy weensy bit bad towards this bloke that they had never met ruining their day. And what about the impact in the room when we eventually got to the rearranged date? Imagine how that might have played out. *'Hi everyone, I'm Neil, and first of all I would like to apologise for not being able to deliver today's material last month. I hope that didn't cause anyone too much inconvenience'* (with a friendly smile on my face). Not the best way to win friends and influence people. Not the best first impression to create.

So, armed with a bucket of Lucozade Lime (Lite!) and a couple of packets of Soothers (I think this may be what is called 'product placement') I arrived. I did the event. I got through the material. There were points in the day when it was obvious that I was struggling with my voice. But somehow I managed for that entire day, in front of my customer, to man up (must have had the hotline open throughout to my inner woman) and just get on with it the best I could.

At certain points throughout the day the event facilitator said to me, *'Neil, I can see you're not feeling that great today, but I just want to say that I really appreciate you being here doing this, as reorganising it would have been a nightmare.'* When the feedback came in, there was the occasional remark about the voice being shaky at times (OK, at one point in the afternoon I had to have an emergency cup of hot tea rushed to my side because the voice was definitely evaporating), but everyone achieved their objectives for attending and everyone learned what they were there to soak in.

There you have it: job done. End of story. Well, not quite. Three months later I was back doing another event for the same client, with a different audience but the same facilitator. We met ahead of the day's proceedings and the first thing he asked was, 'How are you feeling today, Neil?' He referred to the previous event and reiterated how he really felt for me that day and appreciated what I had done, when most others would have made an excuse and cancelled. He then went on to say that he had recalled reading something years earlier, which had stuck with him through his entire working life, and he had associated this with me on 'man-flu Monday'. In short, what he told me was as follows:

> Being a true professional is doing your job at the highest level you are capable of, at all times, without exception, and especially on the days, or the occasions, that quite frankly you would much, much rather be anywhere else than in that room at that time. But still you are there and you are getting the job done at the highest level possible.

I hadn't thought of it in quite that way before, but the 'true professional' he described was someone like many of the successful and respected people I have had the pleasure to encounter throughout my business life. What kind of professional do you want to be?

Danger zones

I have just spent a bit of time giving guidance on what to do in order for you to be a professional in your role. It is clear, therefore, that the things to avoid include ignoring all of the aforementioned. But are there some other common traps that people occasionally fall into, which before they even know it damage their professional credentials? Most definitely there are, and many of these will be covered elsewhere in this book – but there are a couple of points that are better placed here.

It is not your job to cause problems, create barriers and make your manager's life difficult. Your boss may have a team of say six people that they are responsible for. They should be dividing their time and attention fairly evenly across the entire team. If one person is taking up 50 per cent of their quality team time (for example, because of an ongoing performance shortfall), then that means the rest of the team are being short-changed. Of course, even the very best people will not go through their entire career without having mishaps along the way. Unexpected issues arise, or a mistake is made. This is accepted and understood, and hopefully it will all be taken in the wider context of the person's otherwise high standards of professionalism. It is also, in some respects, why you have a boss: so that when something doesn't go to plan or something out of the ordinary occurs there is someone who can resolve issues and give guidance. Nobody expects you never to have a problem; what is important is not to be someone who is by their very nature, in the working environment, argumentative, negative or causing issues that damage team morale.

A good way to think about your relationship with your manager is to look upon it as if you were saving money in a bank account for something that is really valuable and important to you (this analogy is adapted from Covey 2009). Is your career really valuable and important to you? Assuming your answer is yes, then it should be easy to see where I am going with this. All the good stuff you do – the extra hours, the important snippets of information gleaned from a client, the attention to detail that saves the day, the anticipation and prevention of a problem – all of these are deposits you are making into that rainy-day savings account. You don't want to withdraw any money from that account. But once in a blue moon something goes wrong (e.g. a genuine, out-of-character mistake is made) and you may very well need to make a withdrawal. On that day, against the background of all the good work you have done, you hope that the banker (your manager) will allow you that withdrawal without any penalties applying. Now consider your manager's reaction to issues arising if that account is already overdrawn, or worse still if it is just always 'in the red', because you have had lots of 'rainy days'.

Another aspect of professionalism takes us back to this chapter's first definition of a professional, as someone who is paid to do a job. If you are taking your salary at the end of the month it isn't really appropriate at any level to be saying anything negative about your employer, or doing anything that damages your employer's interests, or impacts negatively upon team morale. You are paid to represent your employer in the best light and with a positive emphasis. You are most definitely not paid to damage the brand. Arguably, damaging the business in this way could be construed as misconduct, perhaps even gross misconduct. If you are working somewhere where you feel you can't adhere to this professional credibility then you just have to continue going about your work to a high professional standard without criticising your employer (it is worth noting that quite often people who do criticise their employer also, inadvertently, damage themselves). Either that or go and work for someone else. If you have ever wondered what the awkward interview question *'Why do you want to leave your current job?'* is all about, then there you have a big part of it. I would not seek to employ someone who is prepared to say to complete strangers anything negative about their current employer. It's them being slated today, and us tomorrow: avoid; avoid; avoid.

You must always remember who is paying the bills. Your current employer is the one that has taken you to where you are currently positioned. Do not assume that if you tell a friendly face about an issue you are having at work, and tell that person not to let it go any further, that it won't in some way, later on, cause damage either to your professional credentials or to the reputation of your current employer. There is only one way to guarantee that this doesn't happen, and that is to discuss it with no one other than your employer.

EARLY DAYS – ESTABLISHING YOURSELF IN THE ROLE

The next theme I want to focus on involves those early days when someone has just moved into a new role. This is a particularly intense time for all concerned. First of all there is the pressure the new person feels under as they commence their new role. At the beginning they have no track record, they are an unknown quantity, and they will probably be feeling that they need to demonstrate to themselves and those around them that they are able to perform at the right level. When it comes to the people they are now working alongside, there are two main groups. Firstly, and most importantly, there are those who employed them in the first place, including their new line manager (of course in some instances the employer and the manager may be the same person). The manager/employer will be paying quite a lot of attention to this new person's performance and how they interact with the team. Secondly, there is the team itself. This group will be watching with interest how the new person settles in. Did the recruitment panel make a good choice?

These early days thus create a working environment that is different from the one that will be encountered when someone is more established in the role, fully integrated and accepted as being a 'one of us' team member. What follows are some specific pieces of guidance with these early days in mind.

Wake up – you have work today!

Let's start at the very beginning. Surely no one would ever turn up late on their first day at work, would they? Well, you would be surprised. On the three occasions I have directly witnessed this happen, and where a plausible excuse was given, each of the characters involved proved in a very short time to be a disastrous recruitment decision. Hands up – I was one of those recruitment decision makers. I still shudder at the thought of it. We all sat there waiting one Monday morning, as 10 o'clock drew ever closer, wondering if the person involved was actually coming to work for us. They were supposed to start at 9 o'clock. Perhaps they were going to phone to say they had changed their mind. Eventually they sauntered in and casually explained that they hadn't realised how far our office was from their house (they had been to our office before for an interview), and that they had also got stuck in a traffic jam. How unlucky! Look, there is a crystal-clear message here. DON'T BE LATE.

This leads me to another personal example (Case study 3.2). What is described here, or something very similar, has happened on a number of occasions, so it is by no means an isolated example of inspiration on the part of the individual concerned.

Case study 3.2 Don't be late

A person we employed on a part-time, fixed-term seasonal contract, at Assistant Ecologist level, arrived for their first day at work. This person lived a fair distance away and their journey to work required them to drive along a major city bypass. They knew that the bypass was always busy during rush-hour periods, but had never needed to use it at that time up until now, and they weren't sure how it would impact upon their commute.

On arrival (they were slightly early, as it happened), and after we had carried out the normal formalities for a new team member, I asked about the journey to work that morning. The new recruit explained that they had had no problems whatsoever, and that they had tested the route the previous week by doing a dummy run from their house to our office on Monday at exactly the same time so that they knew precisely, as far as they could control (there is that word again ...), what to expect in terms of traffic and time. Furthermore, the day before (Sunday) they had taken the same journey in the quietest of conditions, to establish what would be involved if they were working a late shift.

In Case study 3.2, the new recruit did not leave things to luck. They practised, and thus gave themselves the best chance of achieving the best result. A perfect example of one of my favourite mantras, as adapted from the quote at the start of Chapter 2: *You don't have to be different to be good, being good is the difference.*

As an employer, when you see a member of your team put so much thought and commitment into the basic stuff, you know that you have made a good recruitment decision. As an employee, get the simple stuff right and make the right impression from the very start, because first impressions can last forever. You might think, *'That's rubbish, they don't.'* But you have just read about my first impressions of someone we employed some time ago (Case study 3.2). Those impressions are still very firmly fixed in my mind, and a big part of what I relate to when I think about this person. Someone who from the moment they first walked through the door had immediately made a deposit into their professional bank account. So go ahead, dismiss it as much as you like. Roll that dice, take that risk. After all, it's only your career that you are gambling with. It's not that important. It's not that valuable. You can always get another one. I will revisit the whole matter of first impressions later in this chapter.

Do the research

Look, listen, ask and learn. To start with, during those early days, you have a *'no such thing as a stupid question'* window of opportunity to ask away and not be thought of badly. Once the curtains have been drawn on that window, it then becomes potentially awkward or embarrassing to expose yourself to the risk of appearing ill-informed, uninterested or unprofessional.

There is certainly some generic information that you need to get acquainted with right from the very start, and most of it will be stuff that you will get away with asking about during those early days. By asking certain questions you may in fact be presenting yourself as being interested and therefore committed to your new role. Suppose you find yourself taking a longish journey in a car with one of the business owners. Why not ask them about their business? Most people who own businesses are proud of what they have achieved and, if asked, are more than keen to tell their story.

Bear in mind, however, that some of the information that could prove useful can quite often be easily found without asking anyone (e.g. by using the internet). So think carefully about what you can find out for yourself (and perhaps have verified later) and what you can only really establish through asking the right person the right questions.

Among the points you should know are the answers to questions such as what is the company structure, who owns the business, who are the directors or partners? How are senior roles and responsibilities split between the key employees? What are the products/services available to customers? Who are the key customers and the key industry sectors the business is involved with? In fact many of these questions could have been asked during the interview phase, but just in case they weren't or you don't remember, get in quickly with understanding all of this stuff.

Other areas that could prove very useful would be an understanding of the company's core values, mission statement and vision. Are these things in place? If they are, do you appreciate precisely what they mean and how they cascade down to your own day-to-day activities within the business? Moving on from these areas, what is the history of the company? How did it get to where it is today? Why is it called what it is called? What are the functions of other departments or roles within the business? What other services are provided beyond ecology? There is so much to find out if you are just interested and inquisitive enough.

I once asked a new recruit, who had been with us a couple of weeks, what they thought about our website, only to receive a blank stare screaming 'swift-like' back at me. I was truly surprised that they hadn't bothered to find out anything about their new employer by means of an obvious and freely available resource. To their mind it wasn't necessary. I think their answer was that they had been busy doing other things. By the end of that day they had not only seen our website but they had given me a critical appraisal as to where it could be improved (which also served to satisfy me that they had looked at it properly). If you are working for someone at the moment and you are not aware of what's on their website, then for goodness sake check it out now. Apart from saving you potential embarrassment, you will probably find lots of useful information that could help you in your everyday work. For example, how often have you spent time describing to a client how to get to your office? How stupid will you feel when the client arrives and tells you that they found the *'How to find us'* link on the company website that you failed to mention? The website tells staff, as well as customers, so much. In all likelihood there will be all kinds of information there about the business: its trading locations, its products, its marketing approach and possibly even details about the customers and projects it has been involved with.

In addition to a website there may very well be a social media presence in the form of a company Linked In and/or Facebook page, for example. Furthermore, does the company have a corporate brochure? You must make yourself familiar with all of these resources. You may find that as well as demonstrating that you are on the ball with regards to these aspects of your employer, it will again help you in your day-to-day business life, as well as perhaps answering some of the questions you might have.

The office – a hazardous habitat

An office environment can be a particularly hazardous habitat, and it's not just the trip hazards I am talking about. Quite frankly you don't know what you might be walking into in terms of your new team members' opinions, allegiances, company history, and so on. In order to protect yourself and help ensure that you haven't inadvertently made your new manager's biggest 'pain in the backside' your first new best friend, there are some simple rules you can follow.

Initially, as far as possible, aim to slowly edge yourself into the new team and begin to understand how this new working habitat operates in terms of its culture and the individuals already there. Almost without exception (please do not go on the premise that you are different) you must do this carefully, without appearing to get ahead of yourself. This could take anything from a few weeks to many months depending on where you are, but it most certainly shouldn't be taking just days. If it does, you are in danger of not having picked up on important things that could potentially trip you up (a slightly different kind of trip hazard). The best approach is to be friendly with everyone, but friends with no one, at least until such time as you fully know what's going on. In contrast to how being friendly works, making enemies most definitely doesn't. I once knew someone who had an open disagreement with a member of staff from another department early on in their new job, only to find out the following day that this person was in fact senior to their own boss (I wonder how thorough their research had been at the outset). Let's just say that some backtracking was required in order to retrieve the situation.

Next, try and just be yourself, but a toned-down version, even more so if you are naturally a loud, extrovert type. You just need to tighten the reins for a short period and gradually become more the 'full you' as the team gets to know you better. And while on the subject of your team members, bear in mind that each of them may be operating at a different pace when it comes to accepting you fully into their environment. So don't write off slow burners in this respect. In fact these people, who are taking longer to let you in, quite often are the best ones to eventually have close working relationships with.

Make an effort with everyone you come across. That includes some people that you may not initially think of as important. I am talking here about those in other departments, car-park attendants, receptionists, cleaners, mail-room staff, building maintenance and the like. I am not suggesting that you spend every day in detailed conversation with these people, but you should introduce yourself, always say good morning and goodnight, and so on. In some business cultures, roles such as these can be treated as 'less professional', and their vital importance to the smooth running of the office can be forgotten. On a slight tangent, I once knew a manager who always asked the receptionist their opinion about individuals who came in for an interview. The receptionist's opinion had influence on his thinking. Overall it's all about building rapport and establishing connections, but doing so with a little bit of thought, bearing in mind what you are potentially walking into.

First impressions

First impressions really do matter, and managers are always keen to see how a new recruit interacts with the team on day one. Naturally they are hoping everyone is happy with their choice of new team member. Therefore, as someone just joining a team, you have to make the right first impressions. We will talk more about that in a minute, I promise.

First of all, however, flipping things around, how often do employers put as much emphasis on the first impressions that they and their team may have on the new member of staff? Let's think about this. Is the team briefed ahead of the person arriving? Is at least one of the people involved in the interview process present that morning (a familiar face) to meet and greet? Is time set aside in the diary that day (not later in the week) for a welcome chat? Does the new recruit get properly introduced to their colleagues? Is their induction scheduled in? Do they get shown where the toilets are so that, if none of the aforementioned happens, they can at least lock themselves away and quietly scream 'What the Gordon Bennett have I done?'

So please, if you are an employer, or an employee already established within a business that is about to have someone new walk through the door, put yourself in that person's shoes. Be kind, be considerate, think about their first impressions of your working environment, and what you and your team can be doing to get things off to a flying start (which is to everyone's benefit), as opposed to a slow stutter.

Now, as promised a minute ago, let's approach the subject from the new employee's point of view. Your first day at a new place of work, no matter how many jobs you have had before, is a stressful situation. For everyone else in the room it's just another day, but for you it's very different. There are so many unknowns. So many questions that may be scrambling your brain. So much pressure to make the right, the very best, first impression. In your head you may even be running through all the disaster scenarios. 'What if no one likes me? What if my car breaks down on the way? Where is the nearest sandwich shop? Am I allowed a lunch break? ... I wish I had never left my last job. After all, I wasn't that unhappy where I was.'

OK – breathe, relax and get your brain in gear. There is so much that you can actually control and have influence over in order to get through this period confidently. We have already talked about being professional, doing the research, avoiding metaphorical trip hazards and so on. Table 3.1 provides a list of some other positive behaviours that can be adopted when starting a new role. In addition to this, the section on *Communicating with your line manager* (later in this chapter) provides some guidance on how you should relate to with those senior to you in the business.

What we have been discussing, if handled properly, all feeds into the first impressions that everyone you will encounter for the first time will have of you. It

Table 3.1 Early days: positive behaviours to adopt

Behaviour	Behaviour in practice
Be punctual and a good time keeper	On the first day, arrive on time or preferably slightly early. Take shorter breaks than you are entitled to. Work beyond your agreed finish time (even if it's just 15 minutes). Stay a good timekeeper for ever more. Don't be a clock watcher. Build a reputation for being hardworking.
Handshake ready	You are going to be doing this a lot. Practise and develop a good handshaking technique: keep your fingers together and your thumb pointing upwards; slide your hand into the other person's hand so that the skin between each of your thumbs and forefingers touch. The handshake should be firm and only last a few seconds. You should shake the other person's hand only once or twice before releasing. Ask a friend for feedback. Your handshake has an impact, every time, on someone's first impressions of you.
Names, roles and contact numbers	If you haven't been given one, ask for an organisational chart showing team members, their reporting lines and their roles in the business. Ensure you have their email addresses and phone extensions/mobile numbers.
Ask questions	You are not expected to know that much about your new role, so it would be odd if you didn't ask questions or ask for advice. Silence in this respect is usually bad news. Asking questions and accepting offers of assistance helps build rapport, increases confidence in your ability, and helps you to get things right.
Watch, listen and learn	Two eyes, two ears and one mouth – use them in that proportion (4 to 1). You don't, as yet, know enough to have strong opinions about anything. Your time will come.
Smart appearance	Find out the normal dress code before your first day and stick to it. Always be presentable: hair combed, with clean and tidy clothes and shoes. Also pay attention to your personal hygiene.
Be prepared for field work, etc.	If you are expected to do field work ensure that you always have your field kit and clothing available (e.g. in the boot of your car or in a rucksack under your desk), to be changed into at the drop of a hat.
Get to know your boss and the team	What is their history, background and areas of expertise? Check out their profiles on the Web, including on social media. Do they have rules for how the team should behave, interact, etc.? Do they have any pet hates? What about their personal lives, their family, their hobbies? Show interest, and people will reciprocate.
Set your own high standards	Always behave professionally. Do not adopt any of the bad habits of those around you (e.g. it may appear acceptable to those around you to be constantly on Facebook, but you know it's wrong). No personal calls. Keep your mobile on silent (without 'vibrate'). Tell friends and family not to call you unless it's an emergency.

all starts now. There is something worth remembering, however. Before you even walk through that door on that first morning someone, at least one person with some sort of power or importance, must already be thinking very highly of you. After all, you were the one that got the job. You are hitting the ground running, you are already ahead on points, and as the new start you will be cut a bit of slack. Despite all of that, however, you don't want to lose ground, unnecessarily, as a result of something silly that needn't have happened had you been a bit more alert to how you might be perceived in those early days. The numerous first impressions you will be creating on all of those around you can have a huge positive impact if you just take your time and think about it.

ESTABLISHED IN THE ROLE

Passion for the job

In the ecology sector we have a huge percentage of people who are very passionate about their roles and the subjects which they are studying as part of their daily working lives. There is therefore more than a good chance that your colleagues, your manager and the owner of the business you are working for are such people. They have probably got to where they are by working hard, putting in the extra hours and staying focused on the big picture – as opposed to counting the minutes to clocking-off time. If you are going to be as successful as them, then you will also need to hold onto your passion in order to keep your motivation truly focused as you work your way through the long hours that will almost certainly lie ahead of you at certain times of the year.

As you perform your role, it is likely that if passion and enthusiasm about what you are doing is absent, or perceived to be absent, it will be noticed at some level. Once noticed, it may very quickly become a concern. If you are seen as not being that committed to the success of your team, or not being enthusiastic about what you are doing and the goals of the business, you can't really complain if your boss isn't particularly enthusiastic about you. In effect you are telling the company that you are going to play it strictly by the book and offer no flexibility when providing them with your service. As an employee you can open that book at any time you choose. Remember, however, that you can't really complain when your boss reacts by opening their version of the same book and acting painstakingly by it whenever you are looking for some flexibility from them. As an employee, it makes a huge amount of sense for you to show enthusiasm and commitment to the role. Conversely as a boss, you should respond in kind to the people in your team who are committed, who are flexible, and who ensure that their outputs are constantly delivered on time and to the right standard.

In our sector it is just the nature of the job that in some cases, or at certain times of the year, people work much longer hours than they get recognised for. Provided these extra hours are being used effectively (and not just because the person is ineffective), then these are the people who should be getting rewarded with time off in lieu, bonuses, better working conditions, additional benefits, bigger salaries and eventually promotions. One thing that I would want to emphasise, however, is that it isn't simply the number of hours that have been worked that should be measured, or that someone should be compensated or rewarded for. The true measurement is of how effectively that person has used their time in order to get the job completed. It just isn't logical to reward inefficiency or ineffectiveness, no matter how much time has been spent. To demonstrate, I am going to use an example (Case study 3.3).

Case study 3.3   Job done. I'm out of here!

John has just spent all of his time today at his desk working on a report that is required by the client before 09.00 hrs tomorrow morning. The report would not normally take a full day to complete, but he has allowed it to drag out. He finishes work promptly at 17.30 hrs, and on his way out he passes the report to his line manager, saying that it's fully completed and ready for the client.

The line manager says *'Thanks, at last we can now get this out to the client.'* Some minutes later, after John has left, the manager opens the report to find that it's full of errors and some important information is missing. It will take someone a good couple of hours to fix it. Another team member (Robert) steps up and works well into the evening to get it sorted and out to the client on time.

At the end of the week, coincidently, both John and Robert feel that they have worked four hours extra. They both come to the manager on Friday morning asking if they can leave early that afternoon. There is no office cover if they are both allowed to leave early.

Having considered the case study, the obvious question is: Who gets to go home early? The answer is equally obvious: Robert. But there is another question: Has John actually worked any extra hours at all that week? What if you take the view that he used the company's time unproductively, potentially putting the business in an embarrassing position and a client at risk. Added to which, he produced poor work and passed it off as fit for purpose, possibly even knowing that in his absence another team member would have to fix it.

At times it isn't easy being in a managerial position. You can't do right for doing wrong. Furthermore, it's not just about the manager, Robert and John. You can 'bet your cotton socks' that whatever the final outcome, someone elsewhere in the team will react or have an opinion (they probably shouldn't, but you would be a poor and deluded manager to think that they won't). In making decisions, a manager should be considering all angles, all benefits and potential risks. And all of this against ensuring that the overall goals of the business are achieved.

Another example that isn't hugely different to the situation describe in Case study 3.3 relates to how people behave when someone else in the team has stepped up in order to help them out of a sticky situation. To my mind it is inconceivable that you would go home and leave the person who has stepped in to help you to fix the problem, sitting there in your mess, working their socks off to help you. That would never happen, would it? Yet I have seen it happen, and I have heard others talk about times when they have seen it happen. If you need assistance to get your job done then you must of course say so. But when that assistance comes galloping over the hill to save your day you must accept that you are still responsible for the result. It is not professional or credible to then wash your hands of it, or walk away thinking *'it's not my problem any more'*. At the very least you should be watching and learning so that next time you can do it yourself. If not that, then you should be supporting the knight in shining armour with cups of tea, by arranging a takeaway – or, even better, by asking if there is something on their desk you could be doing to help them out while they are spending their precious time helping you.

Hard work and long hours

I have worked in quite a few roles across a number of service-sector businesses. In my experience there are people who work long hours (such as overnight or weekends) everywhere. This isn't just an ecology-sector phenomenon. This is, rightly or wrongly, just what driven people, with their customers in mind, do in order to keep those same customers (internal or external) happy, and therefore the business they are working in successful. In doing so they are also indirectly protecting themselves. They are creating the right impressions and giving themselves the best opportunity for career progression, a pay rise or their boss cutting them some slack at a time when they really need it (that bank account comes to mind again). To put it another way, look after the customer and the customer will come back. A happy customer helps protect the team. If you are the reason the customer is happy the team will protect you. It's all part of human nature in the context of operating in a complex social working environment.

Having worked alongside many hardworking people, across a number of sectors, on the one hand I can say to ecologists, you are definitely not alone. On the other hand, just because you are in good company that doesn't mean that we should cast the issue of hard work and long hours to the side as if it's not really that important and shouldn't be discussed. Yes, working long hours at certain times of the year is something that we just have to accept happens, given the seasonal nature of our profession. Of course there are things, to a point, that an employer can do to make things easier. For example, they can employ seasonal staff in order to help with increased workloads. This all seems quite straightforward, and when it works well not only does the employer and their team get assistance, but for those taken on it is an important step towards a full-time permanent position. But even allowing for solutions, business just keeps coming through the door, and it's a brave business owner that turns it away in the knowledge that in a few months things are going to get much quieter.

For many ecological consultancy teams it appears that no matter how many people they employ there just aren't enough hours in the day. As much as certain roles at a more junior level can be filled relatively quickly, it takes months to find and then train the right people for roles that carry more responsibility. This means that those already established within the business with responsibilities beyond field work are under pressure to complete their tasks. Generally speaking it is these people who find themselves under the most pressure as they juggle the seasonally constrained field work and the associated deadlines for completing the required analysis and reporting.

There is a danger in not recognising within yourself the point beyond which you shouldn't go. Yes, working long hours is sometimes a necessity, but it should not be the norm week after week, month after month. At any given moment you need to be able to recognise when you just have to stop and recharge the batteries, both physical and mental. It is certainly not professional behaviour to be running so low on energy that the tasks that you are seeking to complete are just not being dealt with effectively. In such circumstances both you and your manager should rise above the moment and realise the advantages of you taking a proper break and coming back later to perform at a much higher level.

There is, of course, another far more serious aspect arising from working excessive hours without proper breaks. This relates to the health and safety risks that you expose yourself, your colleagues and your employer to. You, and your manager, need to seriously consider why taking proper breaks and using that break time effectively to ensure that you are fully rested is so important. While doing field work you need to be constantly alert to your environment from a health and safety perspective. Potentially the biggest risk of all concerns driving. It is absolutely essential that you ensure you are fit to drive without the risk of tiredness or a lapse in concentration causing something 'unthinkable'. What if you ever find that you are potentially falling asleep at the wheel? No matter what, you must safely park up the vehicle immediately and sleep. Either that or pass the responsibility of driving to someone else. There can be no pride here on your part, and there can be no complaints from your passengers. Safety is paramount, always. If you are not safe you are definitely not behaving professionally in this or in any other instance.

Hitting objectives

In many strong business models the current-year objectives will be set by the people at the very top, and those objectives will be determined by how they contribute towards longer-term plans, the strategy and the vision of the business. By the time you are told what your specific targets are for the coming year, it is probable that everyone's individual targets will collectively add up to a large chunk of what your manager or department is expected to deliver on behalf of the business. Continuing with this thread, is it not a good idea to ensure that as far as possible your team as a whole, including the manager, produces (and hopefully exceeds) what is expected of them? The first big part of this is ensuring that you are performing competently at the specific tasks you are given on a daily basis. Once you are achieving this, what will it mean for you if you have also made a key contribution to your manager's success in achieving the overall desired result? Well, first of all, you are going to be contributing towards your own and your colleagues' job security. Secondly, your manager is going to want to look after you. If you are one of the biggest contributors to that manager's success, what do you think any sensible manager will do? If they are smart, they are going to fight tooth and nail to protect you and keep you in their team.

Maybe you are not convinced. In that case, let's say that you are 'let go' at the end of a contract. Now move things forward a few months and assume that without you there, that manager's figures are beginning to look not as good as they did, and now they are working harder just to stay afloat, let alone sail. Coming back to the present – if you are truly an asset to their performance, then keeping you helps make them look good. Conversely, losing you carries a risk to their own job security, and their own career development. So for goodness' sake don't lose sight of all of this. Anything you can do to help get your manager through their day and achieve their goals can directly or indirectly benefit you. It is the same manager, when the time is right, who is going to be on your side when a promotion opportunity arises.

By now you may be thinking that you don't have the time to do any more than you are currently doing. But even asking the question, *'Is there anything I can help you with today?'* will totally blow some managers' socks off and put you head and shoulders above everyone else in their eyes. Have you ever asked that question? Going further, do you know what your manager's specific objectives are for the following month or the next quarter or the forthcoming year? I am not saying that your manager is going to share with you all of their targets, but if you have half an idea as to what it is they are being tasked to achieve, you can then see far more clearly where you sit in the grand scheme of things, and where you can add value most effectively.

You might still be thinking, *'Nope, I am not doing that. I have more than enough on my plate and I just don't have the time to ask for more.'* I would challenge you and say, *'Find the time.'* All of these tasks that you are carrying out on a day-to-day basis are components of a bigger picture for the business (or, if not, they should be!). You also have a bigger picture. It is about being effective, providing good customer service, making yourself indispensable, achieving goals and furthering your career. Imagine being in a situation where your manager is considering appointing a second in command. Who do you think your manager can best imagine in the role? Perhaps the person who is already handling some of their workload and the person who understands, at least in part, what their responsibilities include? Do you want to be that person, the one that your manager can see is now ready for a step up?

Credibility is king

Be a credible person. What does that actually mean in the context of being a true professional? It can be taken to mean a range of qualities including acting with integrity, being trustworthy and believable, and being dependable and reliable. What kind of work colleague would you rather be? The person that everyone comes to for all of the gossip, or the person that people know they can talk to in confidence? Irrespective of how you have answered that question (I know some people who would want the attention and dialogue that being the office gossip attracts), what kind of person do you think your employer would want you to be? What's the job? Enough said.

You should not be doing anything or saying anything that your manager or employer would not be happy about if they knew. The best way to achieve this is easy – don't do it or say it. That way, you can't be held responsible or be a target if it all eventually comes out (and so often it does). I am now going to discuss a couple of scenarios that hopefully should demonstrate the extremes you should go to in being credible.

Don't ask people to keep secrets and don't expect secrets to be kept. Be the soul of discretion at all times. Don't gossip, or tell tales. If you hear something that your manager or the business need to be made aware of, then it's your duty to your business, your own position and that of your team to make your manager aware. So don't accept, from anyone, anything in confidence that is detrimental to the business, and make it clear to the person you are engaged with in such cases that you cannot be relied upon in this respect. If they don't like the fact that you are credible, they are probably not the kind of person you want to be close to in any case. Your response will therefore deter them from approaching you again. Probably not a bad result all round.

Don't allow yourself to do things that you know are wrong, or for that matter get involved with anything that others are doing that you know is wrong (such as falsifying survey results). As soon as you do this you are putting yourself at extreme risk should it ever be discovered. If you are involved with others then you become embroiled with something that they have over you. The kind of person that would encourage you to do this is precisely the kind of person that will use it against you again and again, and will feed off your fear that your indiscretion will eventually be found out. Thankfully, in our sector this type of behaviour appears to be very rare.

You are responsible for who you are, for why you do what you do, and for how you achieve what you achieve. If you keep doing the same stuff in the same way you will achieve the same results. Accept responsibility for all that goes on in your business life, and if you are not happy with what you see then do something about it. No one else will – it's down to you, and you alone.

Communicating with your line manager

There are a number of things that you would be wise to consider when it comes to communicating with people senior to you in the workplace. First of all, anyone at a more senior level is deemed by the business to have more authority than you. You may not see it that way yourself, but unfortunately that is the way it is. As such, you have to always be respectful of their position, and you should always ensure that you are being seen as helpful and enthusiastic when anyone senior to you is around.

Your line manager is, in almost every respect (barring your partner, your parents and your children), the most important person in your life. They are responsible for deciding what you do, or don't do, for the bulk of the time that you are awake to the world. They can give you challenging workloads, they can dictate the pace of your development, they can decide whether or not to sign off your holiday slip. If you don't see that they have a big impact on your working life (which, if you are full-time, is a big part of your life) then perhaps you need to reassess things. They may choose to give you an easy time and let you off the hook in certain respects. If they do, it most certainly is because they choose to, and they could just as well come in to work tomorrow morning and decide to completely change their approach. Considering all of this, you should look after your manager – and if you do, fingers crossed, they are sensible and business-wise enough to appreciate your efforts and look after you.

When it comes to communication with your line manager or those senior to them, there are some basics that you should adhere to, as listed in Table 3.2.

Table 3.2 Tips for communicating with your line manager

Tip	Description
1	Never interrupt them while they are speaking; allow them to fully explain their point of view.
2	Never disagree with them when others are present (e.g. in the office at their desk, during a team meeting or in front of a customer or supplier).
3	If you have something happening in your life that is going to impact upon your performance or your ability to complete a task then let them know immediately. Don't wait until it's too late for them to manage effectively and then becomes an even bigger headache for someone else, or, worse still, it sounds like an excuse.
4	Respond well to feedback (see Appendix 1).
5	Carefully consider how you phrase what you are trying to say so that it comes across positively. For example, instead of saying 'That's just not achievable' or 'I can't do it', say 'In order to achieve that we need to consider ...' or 'I can do this if I have...' If you are not going to contribute positively, then seriously think about whether you should say anything at all.

Don't ever view your manager as a friend, and if you are a manager please do not view your team members as friends. The best approach is to be friendly, but not friends. You must always remember that what has brought you together is the company's need to have teams of people performing effectively in order to get the job done for its customers. Overall, that is what matters, and if things ever get sticky your manager will not be able to be your friend: not unless they are willing to also risk their own career.

Top ten ongoing positive behaviours

To conclude this chapter I would like to leave you with my top ten positive behaviours that can be adopted when working in a service-sector business environment (Table 3.3). These are behaviours that should be considered in addition to those listed earlier (in Tables 3.1 and 3.2). Taken collectively, these are all within your control. It's up to you whether you take them on board or not.

Table 3.3 Top ten ongoing positive behaviours

Number	Behaviour
1	**Be professional**
	Be credible, dependable and reliable
2	**Support your manager**
	Make your manager's life as easy as possible
3	**Accept responsibility for your own actions**
	Don't make excuses or blame others for your shortcomings
4	**Be honest**
	Don't lie or inflate or hide the truth
5	**Watch, listen, ask and learn**
	2 ears, 2 eyes and 1 mouth: use them collectively in that proportion
6	**Be punctual and a good time keeper**
	Be where you are supposed to be, when you are supposed to be there
7	**Be organised and in control of your environment**
	Tidy desk, tidy email inbox
8	**Always have your diary to hand**
	Do not have separate diaries for work and personal life
9	**Receiving tasks**
	Understand what is being asked and the deadline required
10	**Always have a notebook and pen**
	Be ready to take task instructions and notes effectively

Chapter 4
COMMUNICATION SKILLS

Listen with the intent to understand, not with the intent to reply.

Stephen R Covey (1932–2012)
American educator, author and businessman

Our communication skills play a huge part in us all being effective within a business environment, and for that matter in every other aspect of our lives, from raising children, to our relationships with our partners and our friends, right through to all of the other day-to-day social interactions we have with everyone we engage with. It is not possible within this chapter to cover every aspect of communication thoroughly. Accordingly, my intention is to focus on areas that are more likely to be relevant to your business life on a daily basis. As there are a number of ways to approach the subject, it will be helpful to start with a 'road map' of the areas that will be explored in this chapter and how they relate to each other (Figure 4.1).

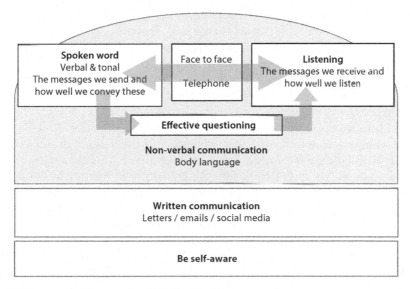

Figure 4.1 Communication – road map to Chapter 4

As shown in Figure 4.1, we are first of all going to discuss *The spoken word*, which includes not only the words we use (the verbal content) but also the manner in which we say things (our tonality, or the way we speak). This will lead on to a discussion of how we *Listen* when others are talking to us. Communication is not just about what has been said; effective communication depends at least as much on how a listener interprets and reacts to what has been said in the first place.

So often, so much of where a conversation goes relates to asking questions, specifically, asking the right questions at the right time. There are a number of questioning styles, as outlined in the section on *Effective questioning*, which can so often influence what is said next (i.e. the response) and how it is said (i.e. the tonality of the response).

Underlying much of this is how we all communicate with each other visually (i.e. in our body language). Body language is also referred to as non-verbal communication, and this is how I have chosen to formally describe it in this chapter. I want to stress here that the section on *Non-verbal communication* relates purely to the visual aspects of how we communicate and how we interpret what the body language of others is telling us (in other words, excluding the verbal and tonal).

The section on *Written communication* focuses purely on the written word. While reading that section, however, please don't lose sight of the subject matter covered elsewhere in the chapter, as a fair amount of what is described throughout could also be considered when writing (e.g. appropriate use of words, effective questioning). Finally, to close off the chapter, we will explore the subject of *Self-awareness*.

The components of communication

Quite often, the way in which others relate to how we communicate can result mainly or entirely from the words we use. For example, someone reading a book, an email, a letter or a text has little or no additional information, beyond the words themselves, when interpreting what is being said. Therefore, in such cases the words used have to be considered carefully in order that the correct meaning is accurately conveyed to the reader. However, once the setting allows the communicator and the receiver also to hear each other (voice to voice), the message usually becomes easier to interpret. In these situations the effectiveness of the communication does not rely purely on the words we use.

Part of how we are understood also relies on how we speak (tonality). Tonality is the overall effect created through a combination of the volume of our voice, our pitch, our articulation, our energy and our speed of speech. This can also be described as how we sound 'vocally'. Do we sound confident or nervous? Are we speaking softly or shouting? Are we talking more slowly than normal, indicating that we are being very careful and thoughtful as to what we are saying? The examples given, with a host of other vocal clues, give the people listening to us so much more information than is conveyed purely by the words that are coming from our mouths.

Finally, when people can not only hear each other but also see each other (face to face), there is a third aspect to how we may be understood. That is the visual messages we give off that also, in addition to the words and the tonality, help people to understand what we are saying and how we are feeling. This final aspect of communication (non-verbal communication) can, in certain circumstances, carry the greatest influence upon interpretation. Note, however, that this aspect of communication requires the people involved to be face to face whilst conversing. That is why, in some circumstances, there can be a huge benefit to conveying certain messages in the physical presence of the other person. Table 4.1 considers the various aspects of communication, as discussed, relative to the setting (i.e. written only, voice to voice, face to face).

Table 4.1 Co-relationship between settings and components of communication

Setting	Written word (e.g. book, email or text)	Voice to voice (e.g. telephone)	Face to face (e.g. meeting)
Components	Words used	Words used Tonality	Words used Tonality Visual clues

It is worth considering the relative influence of the three interpretive components (Figure 4.2), when there is an inconsistency, uncertainty or confusion of meaning on the part of person receiving the message. In this situation it is thought that the words

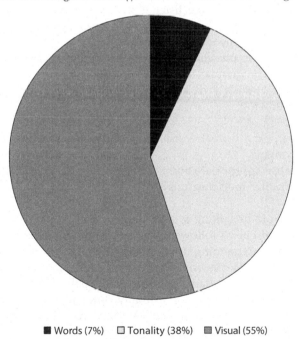

■ Words (7%)   □ Tonality (38%)   ■ Visual (55%)

Figure 4.2 The relative impact of the three components (verbal, tonal and visual) upon interpretation when inconsistencies between the components exist on the part of the recipient (based on Mehrabian 1971, 1972)

themselves become far less important in interpreting the message, carrying a weighting of only 7 per cent in the deciphering of what the communication actually means. The tonality carries considerably higher importance (38 per cent weighting), and the visual signals are most important of all in the eventual interpretation (as high as a 55 per cent weighting). Even where there is no conflict between the different components, it is worth bearing these percentages in mind when we wish to communicate effectively.

THE SPOKEN WORD

We all know how to speak, and most of us do this very well. Making ourselves heard is how we often get on in life. It can be as simple as asking for a train ticket, or ordering a coffee, or discussing where our next holiday is going to be (and what birds are endemic to that part of the world). Whatever scenarios present themselves, we can choose whether or not to communicate our feelings and intentions. Alternatively, maintaining a dignified silence can also, so often, be every bit as effective as speaking.

The words that we use are completely pointless unless there is someone there to listen. Surely the whole point of saying something is in order for someone else to hear what you have said and for them to react in some way. Following this through, therefore, it is the reaction of the listener that really matters and dictates what happens next. Do you keep talking, do you wait for an answer, or do you just shut up and go and speak to someone else instead?

As discussed earlier in this chapter, communication does not rely purely on the words that we use. The manner in which we speak can also have an impact. So, whether it is in a face-to-face or a voice-to-voice setting, we always need to be considering the following:

- Are we using the correct words to describe precisely the message we are trying to convey?
- Is our tonality appropriate and complementary to the words being used in order to reinforce their meaning?

For example, we could be talking to a colleague who has just returned from a site visit. Our words might be as follows: *'That took longer than I expected it would. Was there a problem?'* Now, depending upon the tonality, those words could be spoken with the intent of showing concern for our colleague, or alternatively in a manner that might suggest that they had been inefficient with their time. Each of these approaches would be likely to cause a totally different reaction. On top of this, irrespective of how the person actually reacted, they would certainly be feeling differently internally to each approach.

So it's fairly easy – get the words and the tonality right. Well, as with most things, it is difficult to achieve this at an 'upper-quartile' (i.e. very high) standard all of the time. Sometimes you just blurt something out, or you come across differently to what you intended. Perhaps, even, you have misunderstood the situation and reacted in an inappropriate manner. What then? The best thing to do is just apologise, explain why you said what you did and then start again. It could be phrased something like: *'I am really sorry I said that. I am having a bad day; which is not a reflection on you. I should have been paying more attention. Can we start again?'* My advice would be that whatever you do in this situation you should just accept responsibility for the mistake and take responsibility for moving matters on in the direction it needs to go. Please do not try and bluff your way out of the mess or pretend it was someone else's fault. Credibility is always king!

Your visual behaviours (see *Non-verbal communication,* later in this chapter) will also impact upon how you are interpreted in a face-to-face scenario. For the moment, however, I want to focus on a voice-to-voice setting. This usually means on the phone. It is very important when the people communicating cannot see each other that they are focused both on what they are saying and on their tone, as well as on what they can hear from the other party. There are no visual clues. If you say something with a hint of sarcasm, for example, the receiver won't notice that glint in your eye, or the smile on your face immediately thereafter.

Whether you are on the phone (voice to voice) or speaking face to face, bear in mind that the words you use and the manner in which you say them (i.e. the tonality) are both going to have an impact upon the person you are speaking to, and potentially even an impact upon others in the room who can hear what is being said. The key question is always, are you saying things in a way that is helping you create the reaction or response that is intended? A strong level of self-awareness in this respect will do you no harm whatsoever in enabling you to understand why, on occasions, people may react to you (positively or negatively) the way they do. If you occasionally find that you get reactions that you are not anticipating, you might find the last section of this chapter particularly useful.

Effective telephone technique

A few telephone techniques I have gleaned from others over the years are as follows. Firstly, if you are trying to come across as being friendly and positive then smile while you are conducting the call. Imagine that the person is in front of you and emphasise your visual clues (despite the fact that they can't see you). This helps to ensure that your emotional state is coming over accurately. Next, it's always a good philosophy to speak slowly and make sure that if you are using jargon or putting across detailed information, that you give the other person time to write things down. Repeating names, phone numbers and email addresses is a must. How often do you get passed a message with the wrong name or email address written down? It wastes time, it's not efficient and it's not effective. All because someone either didn't repeat it or the other person didn't read it back, or if either of these things did take place they were rushed over in order to save time. Finally, when taking or making difficult calls, you should, ideally, steal yourself away somewhere quiet, without distractions, so that you can be fully focused on the conversation. Tell your colleagues where you are going and why. Ask that you are not disturbed. Make sure you have a drink with you (not that sort of drink!). Here is another little tip. Try standing up while making a difficult call. I can just see some people's eyes rolling here – 'What a lot of nonsense. There is no way that standing up when making a difficult call or trying to take control of a conversation on the phone ever works.' I was also cynical about this. Someone told me to do it years ago. I did it. I felt a bit awkward the first few times, but it definitely does work and I still do it to this day.

A situation that we all regularly encounter in the office environment is answering a phone and needing to take a message for someone else. When taking messages it is important to repeat back to the caller the information being given. The number of times that a potential hiccup is avoided by doing this is worryingly high, so it makes good sense, as well as being professional. Another problem is that sometimes the caller doesn't want to leave a message, or even a name. Before you know it the caller has gone and you are left wondering who it was and why they were phoning in the first place. Case study 4.1 provides an example of this.

Case study 4.1   Poor telephone answering technique
Setting: The office phone is ringing and John answers it.

John: *Good morning, Echoes Ecology, John speaking. How can I help you?*
Caller: *Good morning. Is Michael in?*
John: *I'm very sorry but Michael is not in today.*
Caller: *That's OK. I'll call again tomorrow.*
John: *Can I ask who's calling, please?*
Caller: *No, you're alright. I'll call back.*
The caller hangs up.

This can then be quite frustrating later on when your colleague comes back and you tell them that someone was looking for them, but you don't know who. You may of course decide to just keep quiet about it. Not necessarily a wise move, as when the caller does eventually get hold of your colleague they may very well mention that they called earlier. Now, you could argue that it wasn't your fault, and the caller didn't give you a chance. Which, to be fair, is true. However, it may be perceived by others that you just weren't that interested because the call was nothing to do with you. Either way, when such an occasion presents itself you can adopt a different approach that is far more effective all round.

In this situation a better approach is to not tell the caller straight away that the person isn't available. Your job is to find out who it is and what they want. After all, there may be someone else, including perhaps yourself, who can assist. You therefore need to take control of the conversation. The way to do this effectively is demonstrated in Case study 4.2.

Case study 4.2   Good telephone answering technique
Setting: The office phone is ringing and Robert answers it.

Robert: *Good morning, Echoes Ecology, Robert speaking. How can I help you?*
Caller: *Good morning. Is Michael in?*
Robert: *Can I ask who is calling please?*
Caller: *Yes, it's Mr Mitchell of Smith and Co.*
Robert: *Mr Mitchell, I am afraid that Michael isn't in at the moment, but I can take a message and ask him to call you back. Alternatively Jane, our senior ecologist, is here, and I am sure she is familiar with your case.*
Mr Mitchell: *Thank you. It was something I specifically wanted to speak to Michael about. So if you could ask him to phone me back at some point later today that would be good.*
Robert: *Of course I can. I will be able to get hold of Michael and ask him to do that. What is the best number for him to reach you on?*
Mr Mitchell: *01234 123456.*
Robert: *Thanks, and can I tell Michael what it is regarding?*
Mr Mitchell: *Yes, it's about the meeting at Smith Street next week. He sent me an email yesterday about it and I just need to check a couple of things with him before I book the meeting room.*
Robert: *I understand, Mr Mitchell. You want Michael to phone you back on 01234 123456, later today, about the email he sent you yesterday regarding the meeting next week at Smith Street.*
Mr Mitchell: *Yes.*
Robert: *I will ensure that he gets your message and that he phones you back. My name is Robert, just in case you need to call about this again. But I assure you I will pass the message on.*
Mr Mitchell: *Thanks, Robert. Much appreciated.*
Robert: *My pleasure. Goodbye, Mr Mitchell.*

In Case study 4.2, you can see how Robert takes control of the conversation right from the start. Notice how Robert attempts twice to see if someone else can help the caller. The first time he gives Mr Mitchell that option it results in Mr Mitchell refusing the opportunity. Later on, by asking Mr Mitchell what the call is about, Robert can potentially reassess whether or not to get Jane involved. In this instance he chooses not to. Also, notice how Robert repeats all of the information, as well as taking ownership of the call and ensuring that Mr Mitchell has his name (given once at the very start, and again at the end of the call).

Thanks and sorry

I want to dedicate some space to two of the most important words in any language. *Thanks* and *sorry*. These words just do not get used often enough in the business world. It's simple. Firstly, be courteous at all times. Secondly, admit it when you have made a mistake or an error of judgement. Here are a couple of examples of when these words are especially appropriate.

> **Example 1 (Thanks)** – A colleague has just spent the last ten minutes of their valuable time, away from their tasks, answering one of your questions and giving you guidance: *'Thanks for that. I really appreciate the time you have taken explaining it all to me.'*
>
> **Example 2 (Sorry)** – It has just occurred to you that a piece of work you produced earlier in the week has an error in it that will impact negatively upon a work colleague or a customer: *'I am sorry about that. It was my responsibility to ensure that it was accurate and it wasn't. I will take full ownership to rectify the situation, as well as to ensure that the same mistake does not occur again.'*

In relation to Example 1, I know some people that struggle with the word 'thanks'. They feel that if someone does something for them within work time, the person is getting paid for doing it, and as such 'thanks' is not necessary. I'm sorry, but we are highly complex social animals that respond and react to our environment. This includes how others respond and react to us. Be nice, say thank you. Make that person know that you do appreciate what they have done for you.

I have just spent a few days with a couple of people new to working within the ecological consultancy sector. They both had many questions throughout our time together. One of them regularly acknowledged and said thanks for my thoughts, and as we parted re-emphasised this to me. The other didn't say thanks during our whole time together. Who do you think made the biggest positive impact upon me? Which of the two do you think I would actually go that extra mile for and help in the future?

In Example 2 (sorry), it may have been someone junior to you who made the mistake which was then missed by yourself when checking, or you chose not to check it. That doesn't matter. You should still be accepting full responsibility for anything that you are tasked with delivering, without passing blame. The most important point is to fix it quickly and learn from it. If indeed someone under your control has let you down, then that's a conversation between you and them on a different day or in a different setting. The fatal words *'It's not my fault'* should be avoided unless genuinely you were absolutely nothing to do with the whole situation in the first place. In that case, it would be very strange if you were being asked about it.

Saying that it wasn't your fault is in effect saying that you don't accept that it was your responsibility and there was nothing you could have done any differently which would have prevented the problem occurring. If it's your responsibility and you were in a position to affect the result, then an apology is most definitely appropriate. Conversely, what about the impact you make if you don't say 'sorry' when you have made an error? You know it's your fault, and for that matter everyone else knows it's your fault. What message does it transmit to the rest of the team? That you don't accept responsibility for your mistakes, that you are not a credible person, that you don't think you did anything wrong, or that you just don't care. Is that the person you want to be perceived as?

Without fail, if you do something wrong, accept responsibility, without suggesting in any way that it is someone else's fault. You then need to take meaningful corrective action to ensure the situation is resolved, repaired or not repeated. An apology is not about you making yourself feel better, it's about making the other person aware that you have made a mistake, accept responsibility and recognise the harm you have done. In a good working environment, having accepted responsibility, you should not be punished for uncharacteristically making a genuine mistake. However, if you do make a mistake, or witness someone else who is about to make a mistake, it would be wholly appropriate to be blamed for trying to cover it up, or not trying to fix it, or not stepping in and preventing your colleague's mistake. Effective behaviour in this respect recognises that 'prevention is better than cure', but when a cure is required, be quick off the mark to take ownership and retrieve the situation.

LISTENING

Listening is just as important as talking. After all, the whole point of talking is in order to be listened to, and when someone else is speaking to you it is because they want you to listen to them. Therefore, does it not make sense, when someone else is speaking, for you to be an excellent listener in order that you can fully understand what is being said? Your understanding of their words allows you to react appropriately, or learn more fully. As with the spoken word, the skill of listening falls broadly into two settings: face to face, and voice to voice.

Face to face

In this scenario both parties not only are able to hear the words used along with the tonality, but also have the benefit of being able to take in information relating to each other's non-verbal communication (body language). For the listener, this means that by and large it is their body language that the speaker will be reacting to, either consciously or subconsciously. Through effective listening (positive active listening: Table 4.2) your 'listening' body language is giving the person who is speaking messages as to whether or not you are paying attention, as well as whether you are reacting positively or negatively to what they are saying. At some point during the process your body language might indicate that you wish to chip in with something relevant or take the conversation in a different direction. In this situation it could merely be a hand gesture or a slight opening of your mouth that sends the '*I want to say something now*' message back to the speaker, causing them to stop speaking and allowing you to put forward your thoughts. Later in this chapter I will go into more detail about body language (see *Non-verbal communication*).

Table 4.2 Positive active listening (PAL)

Positive	Active	Listening
*Consciously take positive steps to pay attention and demonstrate that you are actually listening to what is being said.	Eye contact Nodding or shaking your head Appropriate facial expressions Hand gestures *Occasionally use words or sounds to verify that you are listening (*e.g. hmmm, yeah, I see*). Also use appropriate tonality *Resist the temptation to interrupt or finish sentences or assume where the conversation is going *Give the person your full attention and don't be distracted *Wait until the person has finished saying what they want to say before you start speaking	*Listen with a view to learning more, as opposed to with a view to finding an opportunity to interrupt and therefore deny yourself the opportunity to understand better *Your silence encourages the other person to say more Two ears, two eyes and one mouth: use them in that proportion *Use your ears to listen to what is being said (the words) and how it is being said (the tonality) Use your eyes for visual clues

*To be emphasised in purely voice-to-voice settings

Voice to voice

When speaking with someone over the phone you have a slightly more challenging setting. Now neither party has the benefit of observing visual clues during the exchange of dialogue. It makes complete sense therefore to always be consciously aware of this and adopt a behaviour that allows for it. Therefore, on the phone, the points marked with an asterisk (*) within Table 4.2 should be emphasised and used more often. In effect you have to make more effort in these areas to allow for the missing visual clues that the other person is being denied because they can't see you.

EFFECTIVE QUESTIONING

One method by which we demonstrate we have been listening effectively, and that enables us to develop conversations and ascertain additional information, is through the questions we ask in response to something we have heard or are aware of. Thinking carefully about how we phrase questions is an extremely useful skill to have. A badly phrased question at the wrong time or delivered in an inappropriate manner can take a conversation down a totally different route to that intended, and likewise can raise or lower the temperature if the discussion is of a particularly sensitive nature.

The whole subject of effective questioning is huge, and books have been written on this aspect of communication alone. Here I would like to touch upon some of the more common techniques that are useful to be aware of when engaging with your colleagues, customers or suppliers.

Closed questions

This style of questioning is used when you are looking to close down a particular route, or for a quick answer, expected to be 'yes' or 'no'. In asking a closed question you are encouraging the person you are engaging with to give you no more than the briefest of answers. An example of a closed question would be: *'Can we carry out the badger sett exclusion work next week?'* To which the answer would probably be an enthusiastic 'yes' or an irritated 'no'.

Closed questions are usually better towards the end of an exchange, once the details have been established. Otherwise they carry the risk of stopping the conversation in its tracks and/or causing some degree of frustration for the other person. I recall the case of someone who was trying to strike up a conversation with a colleague from a different location that they had just met. I have changed the details, but it went along the lines shown in Case study 4.3.

Case study 4.3 Poor use of closed questions
Setting: Two colleagues have just met each other for the first time and are driving to a site in order to carry out a badger survey.

Jane: *So are you interested in music, Tom?*
Tom: *No, not really.*
A minute or so passes
Jane: *How long have you worked for the company?*
Tom: *Five years*
Another minute or so passes
Jane: *I heard you have just come back from a trip to Canada. Did you enjoy your holiday?*
Tom: *Yes. It was great.*
Another minute or so passes
Jane: *Have you carried out many badger surveys?*
Tom: *Yes, over 50.*
Another minute or so passes
Jane: *Are you looking forward to this project we have been asked to deliver?*
Tom: *Yes.*
And so things continue uncomfortably like this for some 30 minutes.

At this point I hope that you are beginning to feel the pain that Jane is going through. Tom just isn't opening up. When Jane (not her real name) came to me and described the conversation I had to highlight that she herself was possibly part of the reason why Tom was not being that communicative. I suggested that her closed questioning technique was doing nothing to help Tom to develop a conversation, and by its very nature may even have been discouraging him from wanting to do so.

When are closed questions better placed? Let's try again with another conversation (Case study 4.4). Within this case study the closed questions are shown in **bold** type.

Case study 4.4 Use of closed questions

Setting: Robert has been asked to phone a client regarding a badger sett exclusion that ideally needs to be carried out within the next couple of weeks. His team manager is keen that it takes place next week, as the most experienced team member is on holiday the following week.

Robert: **Good morning, can I speak with Mr Mitchell please?**

Mr Mitchell: *Yes, I am Mr Mitchell, but please call me Bill.*

Robert: Hi Bill. It's Robert here from Echoes Ecology. **Do you have a minute or two to talk about the badger sett exclusion that is required for Smith Street?**

Mr Mitchell: *Yes, of course.*

Robert: *Thanks. As you know we need to work with your site contractor so that they can assist us with doing the work. It shouldn't take more than one day to complete. We have our most experienced person available any day next week and we are hoping that your contractor would also be available. *What is the contractor's availability?*

Mr Mitchell: *As far as I know the contractor is on site all of next week in any case. They are currently planning to work in areas well away from the sett. I think they can help you next week. I can ask them if they are able to move two of their team with the required equipment and materials on to doing this.*

Robert: **Brilliant. Can we go for Monday, please?**

Mr Mitchell: *No, Monday isn't suitable.*

Robert: OK. **Why is Monday not available?*

Mr Mitchell: *Well it is, but not for the full day you need, as the contractors always have a site meeting on a Monday morning. Also there is a specific job required on Monday that I know they will all be busy on for part of the day. Perhaps if you can offer me another day that would be a better approach.*

Robert: *Of course, I understand. OK, can I suggest Tuesday, Thursday or Friday?*

Mr Mitchell: *Thanks. Let's avoid Friday, as they always knock off early in the afternoon. I will ask whether they prefer Tuesday or Thursday and get back to you later today.*

Robert: *Thanks Bill. We'll pencil in those days for the time being. I look forward to hearing from you later today.*
Mr Mitchell: *Thanks Robert. I will phone you back later. Goodbye.*

You may have noticed that during the discussion in Case study 4.4 Robert asked two questions (marked *) that were not closed questions. You may also have noticed that when Mr Mitchell responded to these questions he gave quite a bit more information than when responding to the closed questions. This other questioning style is what is often referred to as 'open' questioning.

Open questions

Open questions are questions that you would normally expect to get much more than just a one-word response to. They are used most effectively in order to encourage the person answering to give you lots of information. There are different groups of open questions, but almost always they incorporate one of the following words:

Why?
What?
How?
Where?
When?
Who?

The first three of these really do encourage the person being asked to provide lots of information, while the final three would or could end up with a fairly specific answer, but not just a 'yes' or a 'no'. These latter three are often used more effectively later on in a conversation when probing for more specific information. In order to demonstrate I would like to go back to Jane and Tom's car journey (Case study 4.3). Let's repeat the scenario, but this time with Jane really thinking about how she is conversing with Tom (Case study 4.5).

Case study 4.5 demonstrates how the whole conversation develops along a totally different line to that in Case study 4.3. There is benefit to both parties, especially Jane, who was possibly worried that the field trip could be awkward. It's a really good approach when meeting people for the first time to break the ice by striking up a conversation with open questions. In fact, by doing this you don't need to say very much yourself. Just ask a good open question, listen effectively, and then ask the next logical open question. It's amazing how people respond. They will talk for ages and feel good that you have been interested enough to ask them about themselves. In fact, not only has Jane managed to get Tom speaking (so that all she has to do is listen), she is also learning about something she is interested in. Bear in mind, however, that if you are the person being asked the questions there comes a point when you should stop talking about yourself and swing the conversation back in the opposite direction.

In Case study 4.5 I have deliberately included different types of open questions, and in Figure 4.3 you can see how all of the questioning styles discussed so far relate to each other. You will notice that an additional style of questioning has been introduced within the figure. This style is referred to as 'probing' questions. These

---

Case study 4.5 Good use of open questions

Setting: Two colleagues have just met each other for the first time and are driving to a site in order to carry out a badger survey.

Jane: *What type of music do you listen to, Tom?*

Tom: *I am not really a 'music' person, but I don't mind if you want to put something on the radio.*

Jane: *Thanks I will maybe do that later. The boss was saying that you have been with the company five years. What were you doing before that?*

Tom: *Yes, I've been here five years. It's been a bit of a rollercoaster, but overall I am really enjoying it. Before all this I used to work for a travel agency.*

Jane: *Goodness. Where in the world did that take you?*

Tom: *Well, nowhere really. I was based in their office in Manchester. But I did get discounted holidays and I was able to go to some marvellous places to watch wildlife. In fact as an ex staff member I still get discounts from them and I am just back from a whale-watching trip to Canada.*

Jane: *Yes, I heard you had just come back from holiday. Can you tell me all about it? I have always wanted to go whale watching.*

Tom: *Well ... ...*

The conversation continues all the way to the site. Jane can't get Tom to shut up about his holiday. Not that she wants him to, as she is already finding out lots of useful information so she can plan her next trip. As for the music – the radio stays off throughout their time together.

---

are used in order to narrow the breadth of the subject matter and guide it along a particular route. For example, in Case study 4.5, Jane's question to Tom *'Can you tell me all about it?'* is very much a probing style of questioning. She has identified something that she wants to explore further, and then narrows down what Tom can talk about by asking a probing question.

Finally, some further examples of how to phrase open questions in order to elicit the greatest amount of information:

Please explain what you mean by ... ?
Please describe more fully ... ?
What are your views about ... ?

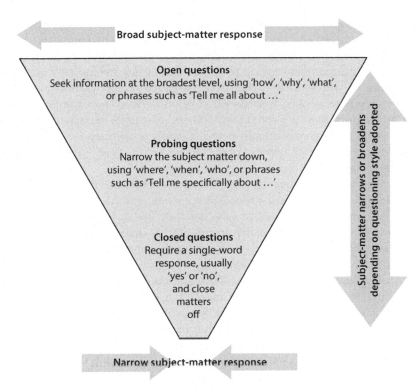

Figure 4.3 Questioning styles relating to breadth of information being sought

NON-VERBAL COMMUNICATION

Human nature

As ecologists, most of us have a good level of understanding about why the natural world behaves as it does. I know ornithologists who can tell me why a bird is behaving in a certain way, and mammalogists who instinctively know where to find a badger sett, or good water vole habitat. Yes, as a group of people we collectively have a huge amount of understanding about the creatures that populate our planet. As such we are probably better placed than many to understand the most complex of mammals that exist in our world: humans, and their behaviour. If humans were wild and lived on remote hillsides or deep oceans and were different to ourselves, as ecologists we would buy books about them, go on courses to learn about them, take expensive overseas trips to observe them in their natural habitat. But no, they are everywhere and they are just like us. Not really as interesting as a rare bird or an elusive cetacean. Not really worth a huge amount of effort. We are one, so there isn't really that much to learn and understand about human behaviour. Or is there?

Humans are highly complex social animals. We communicate not just verbally and vocally, as already described, but through the visual body language we use. In fact, in relation to our senses, body language (also referred to as non-verbal communication) can account for up to 55 per cent of how we portray ourselves to others (Figure 4.2). But more than that, body language also comes into its own when no words have been spoken whatsoever. According to some experts it is said to account for as much as 83 per cent of why others react to us the way they do (Pease and Pease 2004), with what is actually being said only accounting for 11 per cent. The other senses are usually far lower in importance, with smell at 3 per cent, touch at 2 per cent and taste at 1 per cent.

This 83 per cent may seem really high. Surely that cannot be right! But think about it. How often do you just need to look at someone and you can instinctively tell whether they are happy or sad, or whether they are in agreement with you or disgruntled? Our faces tell so many stories. But it doesn't stop there. Even our posture, the positioning of our arms, our legs, our feet and the tilt of our head all give off signals. And then we have the other clues we give off. Our hairstyle, our dress sense, tattoos, piercings, make-up, jewellery and goodness knows what else. These all contribute to the message. Sometimes it's a different message on different occasions. I know people who come to work professionally dressed in suits every day during the week, and at the weekend they are in jeans, leathers and a Motorhead T-shirt. So when you begin to think about how powerful all of these things are collectively, and all of that is without anyone actually saying anything, perhaps now the 83 per cent becomes more understandable.

The long and the short of it is that you would be a fool to think that none of this non-verbal stuff matters. It does. You are being assessed and interpreted all of the time by everyone around you. It may not be right, they may not fully understand you based purely on their visual assessment, but it's nature. It's what we all do, all of the time, either consciously or subconsciously. It's what you do. Now I am not about to suggest you need to change your hairstyle, or what you wear, or whether or not you should have tattoos or piercings, or wear more or less make-up. We are all individuals, and we make decisions every morning about how we present ourselves at work, and about what is deemed professional in each of our working environments. But my message is that you should at least try and avoid anything that could mean that you are perceived negatively or in a less than professional light for the setting you are in. This is because you cannot account for the feelings or thoughts of all customers or suppliers you may encounter during your working day.

You may very well think that it is cool to have a scruffed-up 'edgy' hairstyle. I, as your boss, may very well think that it looks cool as well. But none of that matters. The only thing that does matter during your working day is how you are perceived by those you are engaging with while carrying out your business activities. I can hear it now. *'Well, if someone wants to think that I am disorganised because I have scruffed-up hair, that's their problem, not mine!'* I am afraid you are wrong. You are in a service-sector environment, and your customer's thoughts and feelings are paramount. Lose sight of that, and one day you will lose a customer, or worse still, many customers.

Our ability to communicate enables us to describe our thoughts to others, and conversely to understand the thoughts being described to us. Understanding how to communicate well is therefore a crucial aspect of performing effectively within any social environment, which of course includes a work setting. We spend a huge part of our day in a working environment, and therefore communicating well or communicating badly has a huge impact on how effective we are. As already outlined, that effectiveness depends not just on the words we use and how we say them. So much is also about our corresponding and complementary body language. We have already talked about the importance of using your ears as well as your mouth, and have discussed listening skills, but it does no harm here to emphasise that your positive active listening behaviours are incorporated within the collective body-language signals you are sending to the person who is speaking. Conversely, if you are the one who is talking, paying attention to the listener tells you so much. If you add using your eyes to the equation when you are speaking, then not only are you hearing the reaction, but you are giving yourself the opportunity to see and therefore understand even more precisely how what you are saying is being received.

Using your eyes allows you to focus on the visual signals. For example, is there a smile? How is the person standing? Are their arms open by their sides, or crossed over? What is their random eye movement telling you? Put all of this together, alongside the input from your ears, and you begin to see and hear the whole picture.

Now turn all of this on its head. What about how others perceive your body language and receive the messages you give while communicating? How good is your self-awareness in this respect (see *Be self-aware*, later in this chapter)? Most of us probably feel that we have a really good sense of self-awareness. Sadly, however, the same 'most of us' could be considerably more self-aware a lot of the time.

Table 4.3 describes a selection of common behaviours that can help towards interpreting what someone is really thinking. It is useful to bear these signals in mind both when you are speaking and when you are listening. It is only possible here to touch upon this complex subject very briefly. It is certainly worth taking the time to explore other resources covering this aspect of communication (e.g. Pease and Pease 2004).

Matching, mirroring and pacing

To conclude this section on body language I would like to introduce you to a technique called 'matching, mirroring and pacing'. It is a well-documented approach to communicating effectively with others, whereby you can consciously make an effort to sound and behave, to a point, like the person you are engaging with. Doing this helps to make them feel relaxed about your presence and may result in them engaging with you even more openly. A big part of this approach relates to the fact that most of us are comfortable with people who are similar, in some way, to ourselves. It could be similar interests, similar dress sense or similar backgrounds. As soon as you find that you have something in common with someone you have a connection, and dialogue often becomes much easier. With this approach you are

Table 4.3 Non-verbal communication (body language) – a guide to some common clues

Area of the body	Positioning	Possible meaning
Arms/hands	Crossed	Not receptive
		Defensive
	Hanging by the side	Relaxed
		Open-minded
	Behind head	Confident
	On hips	In control
	Arms extended with hand(s) facing upwards	Inviting
		Asking
	Arms extended with hand(s) pointing	Dictating
		Telling
	Arms extended with hand(s) facing downwards	Supressing
		Taking control
	Joining fingers, like a steeple	Confidence
	Covering mouth with hand	Secretiveness
		Doubt
Legs/feet	Pointing directly at the person they are engaged with	Fully engaged
	Pointing in a direction other than towards the person they are engaged with	Not fully engaged
		Wishing to walk away
Eyes	Looking away	Not trustworthy
		Nervous
	Good eye contact	Credible
		Confident
		Relaxed
Eyebrows	Raised	Surprised
Seated posture	Leaning forward	Interested and fully engaged
	Leaning away	Not fully interested or engaged
Whole body	Mirroring	Fully engaged and relaxed with the person opposite

Health warning – Please bear in mind that any of the behaviours described above, on their own, may not always necessarily mean what is suggested. It is important to always gather information from a variety of clues, including verbal and vocal, before coming to a conclusion about how someone is feeling or what they are thinking.

seeking to focus on particular areas of potential similarity. These areas are language ('matching'), posture ('mirroring') and speed of thought and communication ('pacing').

All of this may at first appear a little sinister. However, you should be aware that it is something that we all do to a certain level, as a subconscious behaviour, in any case. All that is happening through using this technique is that you make a conscious effort to use what is normally a subconscious behaviour more effectively. This then enables you to choose to make minor adjustments to your own behaviour in order to make the other party feel more comfortable in a dialogue with you. Conversely, once you have developed an awareness of the technique you can also read whether or not the opposite party is fully engaged with what you are talking about. Are they either consciously or subconsciously matching, mirroring and pacing you? Table 4.4 explains the technique in more detail.

Table 4.4 Matching, mirroring and pacing defined

Term	Definition	Examples
Matching	Using similar words, language and terms to those of the other person, but avoiding matching their accent. Speaking at a similar volume, to a point (e.g. without going to the level of shouting back at someone).	*If speaking to a customer and they are describing species using their common English names.* You should do the same. If you start using the scientific names, that could have a negative impact on how they relate to you. It may, for example, give them the impression that you think you are better than them, or that you are trying to show off. *If your customer is using certain words that convey how they are feeling.* Occasionally incorporate these words when responding.
Mirroring	Adopting, broadly speaking, a similar body position to that of the other person (imagine that you are their mirror image).	*If you are sitting at a desk having a meeting, and the other person leans forward over the desk.* Doing something similar gives them the impression that you are totally on board and interested in what they are talking about. *If you are both in the mirrored position and you choose to break the mirror.* Could be taken by the other person to mean that you are no longer feeling the same about what is being discussed.
Pacing	Speaking at a similar speed to the other person.	*If your customer is speaking slowly as they are working their way through a problem.* When engaging with them, don't rush. Make a deliberate attempt to explain it at their 'take-in' speed and not at your normal pace.

Health warning – You must be careful using this technique. It is subtle changes that work best (think similar, but not exactly the same), as opposed to copying precisely every aspect of the other person's behaviours, which would be inappropriate to say the least.

WRITTEN COMMUNICATION

Once words are committed to writing you have, in some respects, additional considerations that should be accounted for. Written communication, in a business environment, usually occurs for two main reasons. Firstly, it is a means to document what has been discussed or agreed. It is so important to follow up conversations in writing. In doing so, confusion and misinterpretation are avoided. Secondly, things are committed to writing in order to say something new (when a conversation has not already taken place), or when there are a number of people who need to be given the same message accurately and simultaneously.

Let's go back, briefly, to the first of these scenarios. I don't mean that everything you say to everyone, all of the time, should be followed up in writing. The situations I am talking about relate to anything that is to do with a client, contract fulfilment, allocation of tasks, performance, logistics (meeting time and place) and so on. Therefore, it is these situations, along with the second set of scenarios, that I will discuss, in relation to business-related emails and letters.

It is really interesting to consider the difference between the spoken word and the written word in a business environment. First of all, as soon as something is written it becomes far more real. It is now evidence of what was said (or at least someone's interpretation of what was said) in any given situation. When referring back to a spoken conversation it is all too easy to forget what was said at the time, but as soon as it's in writing the situation is very different. Let's take a fairly straightforward working example to demonstrate this (Case study 4.6).

Case study 4.6   Put it in writing?

A member of staff is speaking to their line manager about another member of the team. They say that they don't feel that the other team member is pulling their weight. The manager says something along the lines of, *'I hear what you are saying, and I can assure you that I am tackling the situation. It will just take a little time for things to become evident to you.'* Now this whole conversation possibly should never have happened, in that it isn't really good form on the manager's part to discuss someone else's performance at this level. Nevertheless these things happen. It's just how, sometimes, we naturally interact.

After a few weeks have passed, the aggrieved team member decides that they can see no improvements from their colleague and they have had enough. They decide to commit their thoughts in writing to their line manager, and worse still they copy in their manager's boss. You now have a totally different situation. Someone, somewhere has to take notice. Something needs to be done about the perceived problem, and it is all going to need to be formalised in order to demonstrate that correct procedure has been followed.

On top of that, the person who has written the email will need to be managed. What about their ongoing relationship with their line manager? And what will that line manager's boss think, both of the manager and of his or her team member? So many things are now occurring because someone has decided to put it all in writing. It can't be taken back, it can't be ignored, and it can't be denied. The evidence is there for all to see.

I am using Case study 4.6 to convey that sometimes you really need to stop and think about when it is and is not appropriate to put things in writing, and also when you should or should not copy in others, and if you do, at what level. There is one thing I do want to emphasise here, however. If you do genuinely have a work-related problem you should never fear speaking to your line manager about it, and then having done so give them the opportunity to manage the solution.

You also need to stop and think about who you are copying in on the emails you send. Case study 4.7 illustrates some aspects of this issue.

Case study 4.7   Think before you 'reply all'

These past few weeks we have been in correspondence with a client about one of their cases. They are working on behalf of a developer (their customer). Therefore, we are two steps removed from the end user. Our client made a mistake and forgot to pass on a vital piece of information to us. They didn't appreciate the potential issue this new information was going to raise, and as per normal practice when they email us the information they copy in the developer. The email includes details about an amended development boundary that apparently has existed for some months. The boundary we are working to is different, and we have already carried out some of the surveys. We quickly speak with our client, to learn that they have known about the change for some months but didn't pass it on to us at the time. Over the phone we are able to agree how matters can be sorted, and in the grand scheme of things everything is retrievable. We write back to our client documenting the phone call and what we have agreed to do in order to ensure that the project is not negatively impacted upon.

When writing back, what are the implications if we copy the developer in to that email? There is nothing to be gained by doing this other than making our client look bad in the eyes of their customer. Therefore, in this instance we break from convention and don't copy their customer in.

Referring to Case study 4.7, how easy would it be to simply press 'reply all' on the email system and accidentally send a copy of that email to everyone? You really have to always be thinking about how people will react to what you are documenting. Sometimes the opposite is true, and you need to copy others in so that everyone is aware of what is happening. Sometimes you don't. Most of the time, thank goodness, it's business as usual.

Let's now explore some positive results relating to following matters up in writing. We have already discussed the importance of this from the point of view of ensuring that the right message has been conveyed to everyone who needs to know what's required or what has been discussed. If you have a conversation with someone this afternoon about something and you then follow it up by email, the other person will probably be appreciative of that. Also, from your point of view it verifies that everything is as you understand it, and it confirms that they agree with what you have written.

This approach is something that I learned the hard way with a boss I had some years ago. I would regularly have him accompany me to meetings with important customers. He was just one of these bosses who always wanted to give the customer everything they wanted, regardless of whether or not we could actually deliver. What tended to happen was that he would agree to something during the client meeting (I was not allowed to contradict him in front of the client), we would then leave the meeting, and he would be totally confident that we could deliver. Some weeks later it would not be uncommon for it to transpire that delivery wasn't possible after all. He would then deny ever having agreed to it in the first place.

For some of my colleagues, in the same situation as myself, this was very stressful. It was a set of circumstances that had prevailed for some time, and at times it led to serious debate within our team. As for me, I just learned to try and dampen down potential promises and manage the client's expectations as best as I could. But I knew that one day I was going to fall foul of circumstances and would end up being blamed for some massive promise that wasn't delivered. In consideration of all this, what I did was simply as follows. After every meeting, I followed up with an email to the client confirming what was discussed and agreed. Before I sent the email I asked my boss to confirm that he was happy with the content and to get back to me with anything he did not agree with. Of course, on the day my boss never did disagree with anything that was written. The meeting had only just taken place and was fresh in his mind, and as far as he was concerned everything was in order, as described. But what this achieved was that it gave me protection from him denying or genuinely forgetting many months later if things did 'hit the fan'. The benefit, all round, in adopting this approach was that I and my boss never had any arguments about anything agreed at a meeting. We had a successful and far less stressful working relationship than would ever have been the case otherwise.

There are so many potential areas for misinterpretation or misunderstanding, or just totally missing something, within our complex world that, quite frankly, you would be crazy not to follow up all of your relevant conversations immediately in writing. Figure 4.4 provides an example of the sort of wording you can use. Here, it is obvious to the recipient that you are documenting the conversation to be helpful,

Dear Mr Mitchell,

Case Number: ABC1234 – Smith Building – Confirmation of Roosting Bats

Thanks for taking my call earlier this afternoon, and for allowing me to put forward our thoughts about the bat roost within the property and how it will affect your development. I thought it would be really useful for you if I put in writing what we discussed and agreed.

My understanding of our discussion is as follows:
- [Point 1 ...]
- [Point 2 ...]
- [Point 3 ...]

If you have any queries please get in touch. Otherwise please confirm that you are happy for us to proceed as described above. We agreed on the phone that you would get back to me no later than Friday morning (30 September).

I now look forward to hearing from you.

Yours sincerely

Robert

Figure 4.4 Conversation followed up in writing

but the indirect benefit to yourself is that you have everything in writing so that it can be referred back to if necessary.

There is lots of information and many resources available covering how to write letters and other forms of written communication. Also, the organisation you work for should already have templates in place for the most common situations, and these should of course include details such as company logo, company registration number, VAT number, trading address(es), website and so on. It's not for me to say how any particular business should portray itself in this respect. I do feel, however, that there are certain rules that can be adhered to when writing either a letter or an email. Table 4.5 puts forward some ideas.

Letter format

It is amazing now to think how the business world I joined 35 years ago managed to get anything done. Business life depended on dictating machines the size of laptops, cardboard folders with cassettes going into the typing room and the letter coming back a day later to be checked over before it was put in an envelope and sent by post to the client. The client would receive the letter the following day (if they were lucky), and would then repeat the process at their end in order to respond. It's not very often these days that we send or receive letters through the post. Email is far faster and more efficient – and any time we do need to produce something in letter

Table 4.5 Letter and email rules

	Letter	Email
Formality	Always formal	Can be formal or less so depending upon nature of relationship between parties
Company name, logo, address, website details, company registration number, VAT number	Everything and always	Everything and always
Heading	Always before content commences	Always before content starts or a proper heading in the subject box, or better still in both places
Salutation	Dear [Name] Dear Mr/Mrs/Miss [Name] Dear Sir/Madam	If person not known then start with 'Dear …' If person is known then can be less formal, and as appropriate to the relationship you have with them (e.g. 'Good morning …', or 'Hi …') Avoid slang or text speak or cool speak like 'Hiya' or 'Hey Dude' Do not start abruptly with just the person's name. Not unless it's your intention to come across angry.
To **Copy in (CC)** **Blind copy in (BCC)**	Not applicable	Do not mix up 'To', 'CC' and 'BCC' boxes: To: is only for people that are expected to deal with the email in some way CC: is for information purposes only BCC: is when you don't want other recipients to see that you have copied in someone else
Content	Keep it formal and professional	Keep it appropriate for the relationship and professional
Signature line	'Yours sincerely' (when you know the name of the person you are writing to) Hand-written or digital signature Your full name and position in business	If person not known then 'Yours sincerely' If relationship is less formal then 'Regards' / 'Kind regards' / 'Best wishes', etc. Your name or digital signature and position in the business
Your contact details (address, email and telephone number)	Always	Always

format, what do we usually do? We type it up, save it as a PDF and attach it to an email.

It is perhaps because letters are so much less common that many of us have forgotten some fairly basic letter-writing skills and rules. When writing a letter, the points mentioned in Table 4.5 should be kept in mind. A letter is definitely still the most formal way of corresponding with someone, and the conventions are still as applicable today as they were before the existence of email.

Email format

If we didn't have email we would still be writing letters, and trust me, if we were still writing letters then many of the emails I see today just would not be acceptable in the business world. Emails are quick letters. They arrive with the recipient in seconds as opposed to days. So we should certainly be thankful for all of this technology and appreciate how easy it makes our lives. I totally get all of this. What I don't get is laziness when emailing a client – poor speech, a bad choice of language and treating a professional relationship in the same way that you would, perhaps, send a message to a pal on Facebook.

In an email, just as in a letter, you must always remember to stay professional and correspond in a professional manner. Yes, you may very well view some of these people as your friends, but your business views them as customers. You are there to represent your business, and there are standards that should be adhered to. This leads me nicely into something that was quite rightly mentioned by Reuben Singleton of Tweed Ecology Ltd when we were discussing professional behaviours relating to the use of email: pressing the 'send' button too quickly and then regretting it later on. We receive an email that, for whatever reason, we don't like, and off we go angrily bashing away at the keyboard in response. Goodness, we have all done it, and usually, a day or so later, we regret what we wrote or the tone we adopted. Either that or we have injected a hint of sarcasm into a response in order to lighten the mood and it has had the total opposite effect. My advice to you would be yes, by all means type that email, especially if it makes you feel better in the moment. But for goodness sake do not send it. At least, not immediately. Save it as a draft, without the recipient's email address included (just in case you subconsciously hit 'send') and let it lie for a day. Then come back to it with fresh eyes and a more balanced perspective. If a day later you still feel the same, then perhaps you should send it after all. The situation may actually warrant your response. Usually, however, when you 'sleep on it' you will see things differently and adopt a calmer approach. The one thing that is certain, however, is once it's sent there is no going back. Therefore, you must think very carefully about the kind of message you are sending on behalf of yourself and your employer.

In situations like this, email may not in fact be the best way to respond. Email (as described earlier in this chapter) is limited, relying only on the words and lacking the tonal and visual aspects of communication. You may well find that it is far better to wait until you have calmed down and then lift the phone and have a sensible discussion with the sender as to how you see the situation, or indeed, if it is that serious, a face-to-face meeting would probably be even more beneficial.

Have you ever been in a situation when you have received an email from someone you don't really know and you are typing a response, but unsure how to style the salutation or sign off? First of all, a good trick is to mirror their approach. So if they have started with 'Dear...' then you reciprocate. If they have signed off 'Yours sincerely' or 'Best wishes' then again you do the same. But there is a line that I would not cross at this stage of the relationship. If they have done something that would not be deemed professionally acceptable for your business, such as starting with 'Hiya, howz it going?' – no, you stick to a professional response (possibly, 'Good afternoon John'). Likewise if they sign off 'bye' – no, stick to a professional response (possibly, 'Best wishes' or 'Regards'). Maybe you are now thinking, *'Old git – how stuffy is he? Life isn't like that anymore. We can be far more relaxed about stuff like this. Everyone is on email. Everyone knows how it works.'* You may well be right; at least for some occasions with some customers and some suppliers. But why take that risk? Why put your professionalism on the line? Why misrepresent the business you work for? The same business that has probably spent many thousands of pounds marketing its brand and conveying its professionalism. You are risking giving the impression that the business is lazy, doesn't have standards, and isn't that professional after all. Perhaps another case study will help (Case study 4.8).

OK, I am exaggerating, more than a bit, in order to make some points. I don't believe for a minute that anyone working within our sector would actually write, in its entirety, what I have shown in Case study 4.8. But I have seen some emails that have, in some respects, not been the forearm's length of a pipistrelle away from some of what is shown. You must always think about the person receiving the email or the letter. In the same way as you need to be conscious of the listener when using the spoken word, what you write needs to take account of the reader. If I was a customer and I got the email shown in Case study 4.8 from a supplier I would be very concerned. Let me elaborate, because there are perhaps some things going on that aren't so obviously wrong.

Firstly, I am not your friend, I am someone you are doing business with. So don't talk to me like I am on a beach in LA about to launch my surfboard. Next, you may be excited about finding the roost. I am less so. This is what I am paying you to do. If you enjoy it so much, then perhaps I should consider paying you less. It doesn't sound like you are working hard work to me. What does BRP mean? And I am going to have to call in the grandchildren to explain btw. A link to your FB (I assume this might mean Facebook) page – I don't think so. You should not be discussing anything about my site or intentions or what's on my land with anyone. Next, costs are not your job? Really? Well as far as I am concerned you need to be the one that is taking ownership of this as you are the person I am dealing with. Do I want to deal with a 'jobsworth'? Who is your boss? Define 'soon' – a day, a week, a month? I am not sure what a *Nyctalus noctula* is (a bat?). Are you trying to come across as if you are better than me because you know some Latin? I don't speak Latin, I speak English; speak to me in my language and use words that I understand. On top of all this we have

---

Case study 4.8   Example of poor email technique
Setting: An ecologist writing to client shortly after discovering a bat roost on site.

To: Client

Cc: Manager; Colleague

Title: We found a roost \o/

Hiya my man. How goes it?

OMG, just thought I should let you know that we found a tree with BRP and hey presto, found a roost within. Totally fab btw, as it's the first time I have ever seen a *Nyctalus noctula* (10 actually). Very exciting! Have loads of pics if you want to see how awesome they are. About to load them up on my FB page [link]. Anyway this means that you won't be able to knock the tree down without a licence. So you will need to put the access track to the site on hold FTTB. Not sure how long it will take to get the paperwork sorted out or how much all that will cost (not my job to work that bit out – my boss will let you know soon).

Bye

John

---

two sites on the go with your company at the moment. Which one does this relate to? Finally, can you please avoid TLA (three letter abbreviations) where I am unlikely to know what you mean? For that matter can you also avoid EFLA (extended four-letter abbreviations) and even LFOSLA (longer five- or six-letter abbreviations). Do you see how irritating that is?

The dreaded email chain

Oh dear, today we have just received an email from a customer in response to a query we sent them several weeks ago. It's all to do with a project where unreasonable deadlines have plagued its progress right from the very start. We have done our absolute best in the time available, and with team members working overnight and weekends. So far we have managed to hit every 'unreasonable' deadline we have been set. We finally have a response to our query, but unfortunately the person writing to us has not deleted the email chain that occurred at their end in the intervening period. I am not about to replicate precisely what was being discussed between our customer and their client (ultimately the people we are all working for on this case). In short, in turns out there is a story behind one of those deadlines. We had a senior member of our team working all of their weekend because the deadline was critical. We had even double-checked this with our client on the Friday. We were told, 'We need it by 10.00 am Monday without question. It's critical we have it by then.' It now turns out, many weeks later, that all of this was in order to give someone in our client's office five days to proofread our report, as they had quite a busy week ahead of them and they were going off on holiday the next again week, which was when it now transpires that the end user actually needed it by. As it happens, the report we submitted before 10.00 am on the Monday as agreed (whoops – there goes someone's personal life) was not looked at until the Thursday of that week. All of this so nicely demonstrated within the chain of emails we have just received.

Now, at the end of the day, it's not really for me to question the background tactics being adopted by our customer here, and it's not unheard of in our sector to see this kind of approach. It's often the people at the sharp end of the delivery that are put under the most pressure. I have no issues with any of that. But here is the point. Having now received this email, and now that the member of staff concerned has seen, in all its glory, what was actually going on, we now have some issues within our own team to deal with. Firstly, the relationship with our customer. Will we ever trust anything they say to us in the future about deadlines? Secondly, and probably more importantly at this moment, the feelings and morale of our individual staff member, as well as our team as a whole. All because someone didn't think to check over what they were about to send us.

The message is quite simple. Do not send out emails without double-checking threads (or chains), and without those threads being directly linked to the message you are giving. It's careless at best, it appears sloppy, and at worst it could really mess things up big time.

Another thing to be conscious of relating to how you communicate with customers and suppliers when sending them business-related emails is as follows. It is all very well having a relaxed approach when speaking to a customer that you have built up a close relationship with, but in the context of writing to them you should still ensure that you keep the tone of the email at a professional level. This was something that Steve Jackson-Matthews picked up on when we spoke about use of email in the workplace. His feeling, quite rightly, was that you must always be thinking one step ahead in these situations. Your client may wish to send your email on elsewhere, and if so will need to decide, before doing so, if what you have written creates the right impression of their 'professional' relationship with the ecologist. Worse still, the customer may, days later, forget about the casual tone of your email and unintentionally forward it. The best approach is to always remain professional when writing to your clients.

On a slightly different note, but with a similar behavioural pattern in mind, have you ever had an email from someone that has a subject matter in the title box that in no way relates to what the email is about? What has probably happened here is that they have found a previous email when the two of you last corresponded, and in their laziness (I am afraid this is the perception that the recipient has) they have simply pressed reply and then started discussing the new subject. At the time it isn't really a big thing because you know what they are talking about. However, once you have filed the email away, and weeks later when you need to find it for some reason, quite frankly you are on a hiding to nothing. You won't remember that the subject line was totally unrelated to the content. Of course you could retitle it for them before you file it away. Brilliant. Someone sends you an email and you have to do their work for them. Don't be that sender. It's sloppy at best, and certainly not effective.

BE SELF-AWARE

In the context of a working environment, self-awareness mostly relates to understanding the potential impact or lack of impact that your actions and your lack of actions has on others around you. This can, at a higher level, be as a result of your normal behaviours as defined by your personality, and of course the personalities of those you are interacting with. Alternatively, it may simply be down to 'out-of-character' behaviours on either side. The people you interact with at work would broadly include your colleagues, your manager, your customers, your suppliers and people you may be encountering on site.

It is important to be self-aware because everyone around you reacts to the things you say and do. Their reaction is usually as a direct result of how you portray yourself in terms of what you say, how you say it, your body language and even down to how you look (e.g. the clothes you are wearing). On the flip side, the whole way that you react to those around you is as a direct result of what they say, how they say it, their body language and how they look.

Here I am suggesting that you develop a raised sense of awareness about yourself. Certain behaviours and approaches within the workplace can have a positive effect on how those around you react to you and your performance levels. Self-awareness is about an appreciation of how you and your behaviours make others feel about you, and why. How others perceive you is their reality. Understanding their reality is extremely useful when trying to communicate, get a point across, develop a debate, negotiate, or interact with them in any other way.

By and large, the way that most people react to and communicate with you is down to how you react to and communicate with them. If someone reacts badly to how you have described something, it is possible that you have caused this reaction. If you had described things in a slightly different way the conversation might have proven to be far more productive. How often have you heard, *'It's not what you said that I have the issue with, it's how you said it'*? Having heard those words, how often do you think to yourself, *'I haven't got a clue what they are talking about'*? However, this isn't the point. The fact is that it doesn't matter what your intentions were when you said it. All that does matter is, did the recipient receive the message correctly? Did what you intended come across effectively? It is how the recipient reacts to you that is the true test of how effectively you have communicated. So you need to be clear in what you are saying, and perhaps also adjust what you are saying in consideration of the other person's communication preference and personality type.

We are going to discuss personality types later in this section, but for the moment let's touch upon communication preferences.

Some people like to talk, while others like to listen more and say less. Some people prefer to communicate by email, while others prefer to make a phone call. Some people like to see diagrams and pictures in order to understand what's being proposed, while others would find it much easier to read about the process involved. And for all of these, there are numerous shades of grey, as well as the context of the situation influencing how we may prefer to do things. I can just picture a person who likes to talk, about to tell a client bad news and deciding to do it all by email. In doing so they are avoiding making that difficult call. But really, if the news is that bad, it's probably far better to make the call and allow the client (the receiver) the opportunity to give an immediate reaction that can then be heard and taken forward from there. In this instance, what do you think most clients would prefer? Someone who is naturally a 'people' person will probably be able to handle the client's reaction well and begin to move things forward far more effectively. Note that the client may like everything in writing in any case (they would be idiots if they didn't). So what to do is make the call, and then follow up the call with an email documenting what has been discussed and agreed.

Some people just switch off when they are faced with a long and detailed email. Others prefer to see everything explained thoroughly. We have had situations in our own business where I have heard someone in our team shout out in frustration, '*It's all been laid out very clearly and thoroughly by email (twice!). Why don't they understand what we need?*' The answer more than likely is that the recipient just doesn't find that format easy to relate to. They might benefit far more from some pictures and a phone call. Again, it would be essential to follow that all up with an email to have it all documented, but the goal is to get the message across effectively in the first place.

You need to fully appreciate that often the reason that people react to you in any particular way, be it positive, neutral or negative, is a reflection on how you are portraying yourself, or a reaction to what you have said or not said, or the way you have said it. The reaction to you will also be influenced by the things you have done or not done, and by the way you go about your daily work. We are all pretty good at reading others around us. We do it all of the time subconsciously, and sometimes consciously. However, most of us unfortunately are nowhere near as competent when it comes to understanding how others perceive ourselves. Having a raised level of self-awareness as to why people react to you in the way they do is a huge strength.

This self-awareness relates to the words you use, the tone you use and the signals that your body language gives off. If only you could see and hear yourself all of the time, you would probably make a number of changes in how you communicate in the widest sense. The simplest way to achieve this is to ask colleagues and friends for feedback on how you have performed during a meeting or a site visit. There is an effective way in which feedback should be given and received (see Appendix 1). Also, the good news is that in this day and age you can increase your self-awareness very easily. We all have phones with digital cameras, laptops and tablets that allow us to take video footage (extreme 'selfie'). Try this yourself when you are making a presentation. It's an eye-opener. Don't be fooled into thinking that you don't really look and sound like that. You do, and unless you are in a very small minority, it will probably make you feel at least a little uncomfortable. This is normal. Treat it as a self-development experience. Lock yourself away somewhere quiet, on your own, and appraise yourself. Is there any feedback you can give yourself that might help you to be that little bit more effective? For example: talk a little slower; a little louder; don't fidget so much; more eye contact. Little improvements can have a massive positive impact upon how you are perceived by others.

Personality profiling

Another thing you can do in order to raise your self-awareness is to explore in a structured way your personality within a working environment. There are numerous well-established and highly regarded methods available (such as Myers Briggs or DISC), enabling you to do this relatively easily.

Some people feel that these personality profiling systems produce results that are not wholly representative of who they are and therefore not accurate. On the one hand I would agree that we are all too complex to be neatly fitted into a predefined box. On the other hand, however, the overall message is often very close to reality for many aspects of your personality. I have to say that in my experience these profile exercises quite often allow people to see aspects of their personalities and behaviours that they may not have been aware of themselves. However, when colleagues are then asked about the perceived flaw in the results, it is not unusual for the onlooker to say, 'Actually, you are like that at times.' So it's important to remember that a large part of the raised awareness achieved through this type of exercise does not relate to how you see yourself. It is how others see you that really matters, and that drives how others interact with you.

I recall a group session where we all did one of these tests. Through our understanding of the various personality profiles we had within the team we were able to see the ways in which we had different working styles and approaches. In light of this we began to understand why different team members would be more likely to be suited towards different aspects of the project. This also gave our line manager a better understanding of what tasks to give to whom, and in some cases who to avoid for certain types of output. Each of us that day would have queried one or two descriptions generated by the profiler about ourselves, but I remember very clearly that all 15 of us felt that the spirit in which we were being described was accurate.

You can carry out these profiling exercises online, and I strongly recommend that you do. It could be an eye-opener in some respects, and having understood yourself in a wider perspective this gives you information against which you can begin to better interact and communicate with others in your working environment. When you answer the questions, don't answer them by describing who you would like to be, answer them based on who you are. It is also useful to place yourself within your working environment when considering the questions posed. Yes, you may very well be a loud extroverted life and soul of any well-oiled pub crawl (most of us are – well at least when I am not having a snooze in the corner), but what are you actually like on a normal work day within your team? (Please don't say 'well-oiled'.) Having gained a better understanding of yourself, it is then possible to explore how someone with your personality type can best communicate, not only with people who are similar to yourself, but also with those of differing personality types.

Chapter 5
ORGANISATIONAL SKILLS

Do what you can, with what you have, where you are.

Theodore Roosevelt (1858–1919)
26th President of the USA, author and explorer

It is going to be extremely difficult to be effective within any workplace if you do not have good organisational skills. As an ecologist you will have numerous tasks on the go during any given day, let alone a week or a month. There will be a number of cases you'll be involved with. Within each of these cases there may be specific component tasks lining themselves up. Add to this the juggling of office time, site time and travelling. Then, as if that wasn't enough, what about the internal non-customer-related tasks that you need to deliver (office administration, expense forms, health and safety risk assessments, meetings, and so on)? It would be easy to be so much better organised if you had less to do.

Yes, we would all be experts in organisation if we only had one thing on the go at any one time. If that were the case nothing else would get in the way. There would be no distractions and no need to prioritise. If you had only one thing to do, you could just crack on and get it done. Back in the real world, that is never the case. Even if you only have one big project to manage, by its very complexity it will contain numerous components and tasks, all of which will be collectively crying out for good organisational skills on your part.

The test of your true organisational ability is when you have loads to do. In that situation how often do you miss things? How often do balls get dropped? How often do you just throw your hands up in the air and think, *'I don't know what to do next'*? The truth behind how organised you really are lies in what happens when you are under a lot of pressure and things begin to get even busier. Are you organised? Do you manage your time effectively? Are you in control?

So, what are organisational skills? What does being organised actually mean? One definition would be as follows:

> Developing skills and techniques in order to create an environment whereby you can maximise the time available to get the right things done in the right order.

Let's take a few moments to really consider why I have defined 'being organised' as I have. *Create an environment* is all about ensuring, as far as is within your control, that your working environment is set up in such a way that you give yourself the best chance of success. *Maximise the time available* is simply about using the time you have to best advantage. Finally, *get the right things done in the right order* is all about prioritisation. I will discuss each of these in turn.

ORGANISING YOUR ENVIRONMENT

First of all, look around your own workplace. Is there anyone you already know that just always seems to be fairly well organised? Someone you can aspire to copy in this respect. How do they do it? What does their working environment look like? What techniques do they use to help them stay in control of their workload, as opposed to having their workload control them? The answers to these questions may not be obvious to you. Having identified such a person, you could quite simply ask them for their hints and tips on the matter. Ask for some time in their diary to provide you with some thoughts on how they organise their lives. The chances are, if they are truly organised, that they will be able to relatively easily fit you in at some point for such a session. They will probably be more than happy that you asked, and will be willing to share as much as possible with you. Putting that idea to the side for the time being, what skills and techniques can I introduce you to, in your quest to become better organised?

In the first instance, however, let's consider what being disorganised looks like. Some typical disorganised behaviours and their consequences are described in Table 5.1.

I could describe many more, but the behaviours listed in Table 5.1 are enough to demonstrate that being disorganised, or less organised than you can be, impacts upon how others view you and, amongst other things, results in poor decision making, lack of focus and wasted time. In fact, lots of wasted time. This means that the precious time you have available to achieve a task is diminished. Less time means there is a risk that the quality of your work will suffer or you will miss a deadline. Worse still, both. Having now exposed yourself to this risk, an indirect, but very real, consequence also lies in the added stress levels. And when you are stressed you are not operating as effectively as you would otherwise be. Things move on, and then you have the dreaded discussion with your line manager about failing to deliver on the particular task. The whole cycle starts again: more time lost, less time to devote to the next task, more stress, and so on.

In the previous paragraph, I have deliberately said '*Having now exposed yourself to this risk ...*', because in many respects you have the ability to be better organised and to reduce the potential for problems occurring. You can choose to make positive changes accordingly, or you can choose to ignore all of this and keep being the person that you always have been. An appropriate expression in this respect comes to mind. '*If you always do what you have always done, you will always get what you always got*' (Henry Ford, founder of Ford Motor Company). In other words, in order to get a different result you need to take a different approach.

Table 5.1 Examples of disorganised behaviours and consequences

Behaviour	Consequence
Desk is a mess	Can't find things quickly
	Poor professional image
	Potential for things getting lost (e.g. put into wrong file)
	Difficult to focus on one thing at a time
Doesn't use diary properly	Misses appointments
	Double-books appointments
	Can't plan productive time properly
	Deadlines occur unexpectedly
	Wastes time before being able to make commitments
Email inbox is full of current and historical mail, a lot of which has already been dealt with and is of no current interest or value	Constantly faced with deciding what to focus on next
	Spends time re-reading stuff
	Cluttered mind
	Needlessly opening/closing emails that no longer matter
	Misses something that should have been responded to weeks ago
Regularly unable to find things	Wastes own time
	Wastes other people's time
	Impacts upon punctuality
	Frustration and stress levels heightened
Turns up on site without correct PPE or right survey equipment	Wastes time
	Extra health and safety exposure due to extra journey
	Embarrasses employer

I am going to now explore some specific examples of where you can create an environment in which you will be better organised, and therefore reduce the potential for wasting time and consequently reduce your stress levels. I am not saying that doing all of this stuff is going to make your life stress-free. I am saying it is going to make things much better than they are currently.

Know where stuff is

If you put your survey equipment and support items away in the right places, and you have documentation all filed in the proper folders (electronic and paper), you immediately have the confidence and clarity in your mind as to where to find them whenever you need them. Part of this lies in having firstly identified and created the places where everything goes. Then making the effort and taking the necessary action to put things quickly away in their proper place. Occasionally, you may come across something that doesn't lend itself to one of your cupboards, filing cabinets, electronic folders or drawers. Now is the time to create a new space, and so on. I cannot impress upon you enough how much time this saves, as well as having the added bonus that you never have to worry about where you have put everything (one less thing for you to think about). You should just instinctively know where everything is. The more you do this the quicker it becomes a subconscious competence (see Chapter 1).

I will give you a straightforward example. You need to phone someone next Friday. You have a choice. You can rely on your memory, you can scribble a note on a piece of paper, you can set up an electronic reminder, or you can write a note in your diary. With the first two solutions you run the risk that you may not remember or that the scrap of paper gets lost amongst the other stuff on your desk. If you put the note in your diary, assuming you use your diary properly (see later) not only will the note be safe, but it will remind you at the right time, and if you were smart you will have even noted the telephone number in your diary at the same time so you don't need to go searching for it. In the meantime your mind has been freed from the burden of having to remember and your desk has been spared another random scrap of paper.

This may sound boring, but I have the same allotted places in my regular environments for all sorts of items. My keys, my wallet, my mobile all have the places where they go when I am at home, at work, in my car, or doing a survey. And when something, very rarely, isn't where it should be it's obvious and I have to immediately find it and put it where it belongs. Boring? I don't think so. What is truly boring and extremely time-consuming, as well as stressful, is searching your house and all of your jacket pockets trying to work out which of the random places your random disorganised behaviours have put the keys this time.

This approach doesn't just lend itself to everyday items like wallets and mobile phones. It applies to everything. Client files in the filing cabinet, emails in the correct electronic folder, phone messages in one place on your desk. If you know where stuff is it will help to unclutter your mind from all of these distractions that don't need to be there, and that are getting in the way of what you should be spending your time thinking about.

Desk environment

Your desk needs to be organised and tidy. I definitely do not subscribe to the view that having lots of unnecessary files on the desk creates an environment that is productive. You may be one of those rare people that isn't affected by this, but most people whose desks are untidy give the impression that they are habitual procrastinators, badly disorganised, or just plain lazy. And more than occasionally that impression is correct. Let's say that you are a 'rare' one, and you can actually work effectively in this type of environment. Well, all of those around you will be interpreting matters quite differently. They will not see that you are organised and in control. So, if you don't do it for yourself, then do it in order to create the right impression to those around you, including your line manager. Apart from this, those around you will treat your work area the way that you treat it. If your desk looks like a dumping ground for dead batteries, guess where everyone else will dump their batteries. If your desk has a month's accumulation of completed files that you haven't put away, it becomes part of the office filing system for those around you to search through every time they can't find a file. Either that, or you are consistently having to demonstrate that the missing file isn't on your desk after all. More time wasted and more frustration. Your desk is most definitely a reflection of the way you work.

Where you put items on your desk can help make life just that little bit easier. Is your phone on the side of you that you would normally answer it from? Are your pens and pencils on the most convenient side? Is your PC positioned so that you can comfortably use it, but maximising the desk space you have for opening out files? Is the desk space orientated according to whether you are right- or left-handed?

What about beyond your desk? Do you have a guest chair? If you do, is it encouraging people to sit next to you for longer than is really necessary? Quite often you will have options regarding how your working area is set up. Have you thought about it? Have you considered having the files you are working on in one place, and the material you have completed on the guest chair (ready to be filed later that day) as a deterrent to anyone who might sit down and take up your time when you can't afford it? I am not saying that you should not be approachable; the point is that you should be dealing with such approaches as effectively as you can.

Do you have your diary easily to hand so you are not taking it out of your rucksack or bag every time you need to refer to it, and then putting it away again? Is your mobile phone accessible but out of immediate line of sight so you are not distracted by social media posts every five minutes? I can't give you a comprehensive list of everything here. The message is to just look around you and think. What things distract you? What things are you always needing that are cumbersome to reach? What is there around you that gets in the way and doesn't help you to be organised or effective?

Consider the people nearby. Every time you make eye contact with one particular person, does it develop into a five-minute chat? Perhaps they don't have the same pressure or workload that you have. Perhaps they can afford that time. Can you? A large potted plant placed in the right position could save you hours of distraction. Brilliant – I can now see lots of trips to garden centres up and down the country as we all create our office jungles. Well, I suppose that's better than a multi-storey pile of stacked bat poo pots. The main point is to think about it, and make the changes. Take control. With the exception of those more senior to you, you decide when you are available for a debate, not someone else.

Let's develop the 'visitor to your desk' scene further by way of an example (Case study 5.1). Bear in mind that what I am about to describe is all with a view to you creating a physical environment and an approach to your work whereby you reduce wasted time.

Case study 5.1   Managing people coming to your desk

Jane is a senior ecologist and regularly needs to find time to proofread reports produced by ecologists working within the business. She is also the point of referral for the team and needs to be accessible to answer technical queries. On a day-to-day basis she feels that she is constantly being distracted from the proofreads, which in turn is making that aspect of her role less effective, as it takes longer and there is the potential that she may miss an error in a report.

She has a spare chair next to her desk for visitors to her workstation. The bulk of the time it is ecologists and assistant ecologists asking her about their case work. Occasionally, her line manager sits with her while discussing complex matters, although usually her meetings with her line manager are either at the line manager's desk or in the meeting room.

Having sat down and properly assessed the situation, Jane has decided to make some changes to how she manages her environment. This is with a view to taking better control of her working environment, but at the same time still being able to deliver what is required of her.

(i) She is now going to place items on her guest chair. This will be a combination of completed files, ready to be filed away, and her 'take-home' reading material. These things can be moved when necessary (e.g. if her line manager needs to discuss something) to the top of her filing cabinet (where she has up until now been putting them), but while they are on the chair they will act as a barrier to someone just coming along and sitting down.

(ii) From now on when someone comes over to her desk she will stand up to speak to them. This will result in conversations being quicker, as the other person will read from her body language that she is about to go somewhere else. This will also act as a deterrent to the visitor sitting down.

(iii) She has just announced to the ecologists that she is going to adopt a surgery-type approach to their queries. She has allotted two 30-minute slots each day when she will be able to discuss their case queries with them. When someone comes to her desk outside of these times she will ask them to come back at the next allotted surgery time. Any emails from the ecologist team that can be dealt with during the surgery time will also be handled at that point in the day. So often these emails involve having a conversation in any case. This is how queries will be dealt with from now on, with the only exception being those that truly require the most urgent of responses.

(iv) At the end of each day she will pass her completed files to an assistant ecologist with ongoing instructions that these files must be put away in the correct place before 09.15 hrs the following morning.

(v) While proofreading or doing other complex work, she is going to switch her mobile phone to silent, and she is going to switch off her automated email desktop alerts. This will help to reduce distractions.

Case study 5.1 hopefully gives you some ideas as to how you can make small changes that still allow you to carry out your role, but in a more controlled, more organised manner. Developing this further, if you have a particularly busy day ahead of you, or if you need time without interruption or distraction, it makes sense for you to advise your colleagues about this at the start of the day. It's also worth pointing out that during such periods it would be inappropriate, having asked your colleagues not to disturb you, to have your mobile at your desk and to be taking personal calls or reading text messages. Not really a credible behaviour, and in any case taking personal calls is not something that should be happening on a regular basis at work.

Sometimes it pays dividends to arrange to work away from your normal desk environment. In particular this can be helpful when you really need to concentrate over a longer period of time and on a more complex piece of work. Most office environments have the ability to offer this (for example, there might be a meeting room that doesn't get used that often), and if such facilities are available it is definitely beneficial to use them. Just ask your line manager if it would be possible for you to do this. Once your manager has agreed to it, then most importantly also tell your colleagues what you are doing and why, so that they don't disturb you. If there isn't a facility at your work premises for this, then occasionally working from home (assuming your home environment is conducive to quiet uninterrupted working) or pitching up at a local café or hotel could work in a similar way. Just be careful in public environments not to leave files open in places where people can see what you are working on, and also not to use your company's name or that of any of your clients while on the phone. When I was working for a large company a few years ago their head office received a complaint from a member of the public about an employee who was heard regularly on his mobile phone while on a long train journey. The member of the public who complained was able to cite the person's name as well as the content of some of his conversations. The investigation and ramifications thereafter were interesting, to say the least.

IT environment

If your desk is a reflection of the way you work, then your PC desktop and your email inbox are a similar reflection. In fact everything that appears on your PC every time you switch it on is an extension of your desk. Therefore, it needs to be considered in the same way. A disorganised inbox, or a folder system not being used properly, or files saved in the wrong place, are a recipe for wasted time and inefficiency.

It shocks me at times to see the state of people's inboxes on their email system. I guess it's easier for people of my generation who first experienced office life long before computers were an everyday working tool. We used to have much of the mail that exists in people's inboxes today physically arriving on our desks each morning. It would not be unusual to be given 25 items of post in one batch with the expectation that it would all be dealt with by the same time the following day (when the next post was due to arrive). In that environment it needed to be dealt with, and if it wasn't it was obvious to everyone around you that you were either extremely busy or struggling. Of course, those two things often happen together. Put simply, the inbox on your PC replicates what used to be items of post physically on your desk, and broadly speaking these items fall into one of four categories (or boxes) as shown in Figure 5.1.

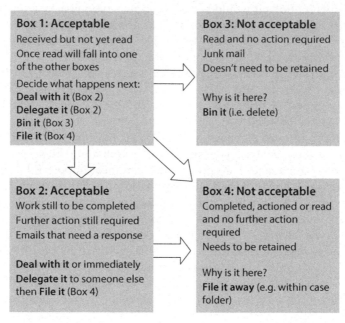

Figure 5.1 What's in your inbox? The four boxes in fact reflect categories of emails within a single inbox

In short, with all emails (and paper post, for that matter) you should immediately decide whether to deal with it yourself, and if so plan it into your workload, or delegate it, and if so do this immediately with strong instructions as to what needs to be done and by when (see Appendix 2, *Effective allocation of tasks*). It's not moving forward until you have passed it on. Once something has been dealt with, then immediately file it away in the right place (i.e. somewhere that you and everyone else will be able to find it again if required to do so). Finally, if the email is of no interest and there is no need to have it, then simply, immediately, bin it.

Ideally one touch per email should result in something happening then and there. By 'one touch' I mean you open it and deal with it then and there. I know that this is not always going to be achievable, but if you make a conscious effort to behave this way it at least gets you in the correct state of mind from the outset. Opening and closing emails repeatedly throughout the day achieves very little. I have just opened an email from CIEEM about my membership renewal. I opened up the document to see how much was due. There was an attachment, which I then opened. I said to myself, '*I will deal with that later,*' and then closed off the documents. Two hours later

I opened the same email and attachment again in order to print off and pay the the the renewal invoice. Idiot! It would have been quicker to just press 'print' the first time, place the invoice on the 'to do' pile on my desk and then file the email away in my CIEEM folder.

Your inbox is no different to your desk when others see it and draw conclusions about how organised you are. If they see your inbox full of historical items, how does your line manager, for example, know what you have actually dealt with and what you have still to respond to? Again, like an untidy desk, it is a poor reflection of your professional image, and it certainly causes time to be wasted and frustrations to occur. In short, at the end of every day you don't really have an excuse for having anything in your inbox as categorised by Box 3 or Box 4 in Figure 5.1. Simply take five minutes and either file them all away or delete them. Once you get into the habit of this it will be much less of an effort. It will be less stressful for you to look at, and you will save time and reduce the risk that you might forget to respond to someone. Alternatively, what I tend to do when I come in first thing is clear out my inbox of anything that shouldn't be there for whatever reason. In fact, because emails can be forwarded to my tablet and my mobile, quite often when I am away from the office I find it easy to keep on top of this kind of work (i.e. stuff that doesn't need much thought and can be dealt with quickly).

There are so many features now available on your PC that can make your life more organised and easier. You can have reminder systems in place, Excel spreadsheets developed for scheduling, project-planning software, calendars and countless other things. Quite frankly, in today's world if you can think about what it is that you need in order to become more organised and effective, it's available. You just need to establish what it is, and then ask someone to create it or buy it. You might think that your boss won't want to spend the money. And yes, in some businesses every pound is a prisoner. But time itself is expensive. If you can demonstrate the case for the expenditure against the time saved and your argument stacks up, then in effect the desired result will save money. How many business owners are going to ignore an opportunity to do that? It's all about thinking through a cost–benefit analysis to ensure and demonstrate that what you are suggesting is of benefit to the business as well as to yourself.

The phone

Where would we be without the telephone? It allows you to attempt to speak to anyone anywhere at a time that suits you. On the receiving end, however, the darn thing just doesn't stop ringing and is a distraction from what you are doing at any given moment. You have the next hour of your life planned, the phone rings, and all of a sudden your time is being eaten into, and quite possibly your priorities now need to change. This is not a recent phenomenon. Office life was like this before mobiles existed. However, the key difference back then was that it was only in your office environment where this occurred. Add mobiles to the equation and there is no getting away from it. And when it's not ringing, you are getting texts and message alerts and social media posts. All of this is wonderful. The amount of technology we have at our fingertips is a world away from what we had even a decade ago.

It can all be used to our advantage. We can quickly find people's numbers (when last did you pick up a *Yellow Pages*?). We can keep in touch with activities elsewhere in the business. We can send pictures from a survey site to the office for a second opinion. We can find our way home when we are lost. All of that, and so much more. But at the same time there is a cost to having all of this, and that relates to the fact that every other beggar has it as well. You are now contactable almost anywhere, at any time, by anyone on the planet.

You need to appreciate that there is a tipping point where the benefits of technology cross over to the disadvantages. In terms of your organisational ability, the negative aspects are the potential for interruption, distraction and deviation away from your tasks and the goals that you have set yourself for that hour or day. So, in the same way that we can take control of other aspects of our office environment, can we adapt our phone-related behaviour?

First of all, keep all of your contacts up to date on the device's directory. As new people phone you, save them as contacts immediately. You will never know when you might need to get hold of them or they might call you again. Having their name appear when they do so is very valuable for putting you in the right frame of mind for speaking to them before you answer the call. It is particularly useful, for example, when doing ECoW-type work, to have all of the relevant contractors' names, responsibilities and numbers saved. Also, have all of your colleagues' numbers saved there, including those running the business, those from other departments, part-time staff, etc. And most importantly have all of the contacts data backed up on the internet, so that if you need to change phones or your phone gets lost, it is one less thing you need to find the time to do. For so many reasons the data on your phone, and indeed any information relating to your network of contacts, is one of the most valuable things that you possess. Look after it and protect it from being lost. It will take a huge amount of effort to recreate it.

Next, definitely have a personal greeting on your voicemail, as opposed to one of those horrible electronic ones. It's so important, when someone phones you and you aren't able to take the call, that they at least know they have dialled the right number, and they are about to leave a message on the right person's phone. It's comical, but I was once with someone who was getting messages left on their answerphone service that were not intended for them. They moaned about this for a couple of days. They were also of the opinion that they never answered their phone unless they recognised the caller. If they had a proper message on their answer machine they would have saved lots of time and frustration, not only for themselves but also for the poor caller who was trying to get hold of someone else altogether.

Should you have two phones? That's a tricky one, and at one point I so nearly deleted the following paragraphs rather than go here. I am going to try and be as diplomatic as I can about this whole subject. I can see why this may occur in certain businesses and why it can make sense to have a separate business phone that you can switch off at night and at the weekends. It can also still be the situation, depending upon someone's personal mobile tariff, that making business-related calls and texts could increase their personal phone bill. Often it is now the case, however, that mobile phone service providers throw competitive text and call plans at us that most people would never come anywhere close to using up, meaning that making business calls does not impact upon our bills. Anything that follows, in favour of a one-phone approach, is on the basis that the employee is not financially disadvantaged.

I am not going to say categorically that you shouldn't have two phones, provided there is a really good reason why it needs to be this way and, if you do, that it is all managed properly. But for many people two phones can be more hassle than they are worth. It's twice as much to carry, to remember to take with you, to find if you have misplaced one of them, and to answer if they both ring at the same time. Yes, having your personal phone and using it for business may mean that you are still contactable in the evening or at the weekend. But if your boss really needs to speak to you desperately then they are going to get hold of you anyway. Also, whether you have a business or personal mobile, if the person is already stored as a contact you will see who is calling and you could elect not to answer it in any case. Out of working hours, that is always your shout.

I am all for making life as easy and efficient as possible. It's like having two diaries (one for work and one for personal stuff) – quite often you end up just making life more complicated in order to demonstrate a 'once in a blue moon' point. I am just not always convinced that the point you might think you are demonstrating is worth the hassle at the end of the day. I am assuming, of course, that you are not regularly receiving 'out of hours' calls, and that when you do get them they are about things that make life easier. By way of example – getting a call in the evening to tell you that you don't need to get up at 3 o'clock tomorrow morning because the bat survey has been cancelled. Is that a call you would rather take, or would you rather find out in the morning when you switch your work mobile back on, having been awakened by the alarm on your personal phone? You see, for an ecologist it's just not that easy to separate working life from personal life so neatly.

I can also see it from the other point of view. When you are not at work you don't want to be getting work-related calls. Why should you? You are on holiday, relaxing with friends and family. You don't want that client phoning you about the problems they are having at their site. There isn't that much you can do about it anyway. But let's think about what happens if they phone the work mobile that you left at home. Did you change the voice message to say you were on holiday, and who to contact in your absence? If not, the customer is now leaving a message on that phone expecting you to call back. Eventually, in two weeks, you will get the message that has been left – or more likely the messages. In summary, if you are someone who is going down the two-phones route then you need to ensure that you manage it properly and in consideration of your customers and colleagues.

You also have lots of other options on your mobile phone that you can use to make your life more organised. You can, for example, switch on/off social media alerts, email alerts, text notifications. In addition you can have different ring tones for different types of caller. During a business day switch off the social media alerts, and so on. You really don't want to be seen as regularly being distracted by personal correspondence in your working environment, and on top of all that you have work commitments to deliver. Imagine telling your line manager that the report is late, when that same line manager has watched you texting your partner about next year's holiday all morning.

If you are left a message to phone someone back, in effect this is a piece of mail. As such you could deal with it in a similar way to how I have described your email inbox. Take time out during the day to make all of those quick calls back to people. If you don't need to call them back, drop them an email or a text with the answer they are looking for. Prioritise who to call first, and so on. If you answer a call and it's not convenient to talk to that person at that moment, ask them if you can call back later. Agree when you are going to call them back, and, extremely importantly, honour that commitment. To forget or later decide not to bother is not professional, is not service-orientated and certainly does nothing to enhance your credibility. Treat people the way you would want to be treated yourself.

Diary management

We have so many options today for keeping ourselves organised and making sure we are at the right place at the right time. We can have calendars and reminder systems on our PCs, our mobile phones, our tablets and of course the good old-fashioned paper diary.

First of all, if you don't have a diary system of some sort then you are setting yourself up to fail to deliver in some respect, and consequently to disappoint those around you. No matter what level you are at you will need to have a prompt about things to do: site visits, when to phone someone back, what day the team meeting is happening, and so on. Whether it's electronic or paper doesn't really matter, as long as it's something. Personally I prefer paper, week to view. Paper doesn't run out of battery life, and paper can be used during a meeting without switching a phone on and risking the distraction of an incoming call or text.

Having a diary in place and religiously populating it with everything you need to be prompted about is only part of the story, however. You must also remember to look at it regularly so that you know what's coming up tomorrow, by the end of the week, the following week and so on. There is no point keeping a diary if you forget to check what's in it, and checking tomorrow morning when you arrive at work is sometimes just going to be too late. You get into the office at 08.45 hrs to find that you should be heading to a meeting 90 minutes away that is due to start at 09.00 hrs. The time to look at your diary is the day before, and the week before. Having a separate work-related diary that stays in the office, or that you have forgotten to take home with you, is risky.

As with the 'two phones or one?' question, there are different views on this. Whatever approach you take there can be positive or negative results depending upon the circumstances of the moment. I suggest that you should only have one place where you put everything, both personal and professional. I can hear some people objecting to this: '*I have no intention of letting my work interfere with my personal life.*' Well, if that's how you really feel, then that's even more of a reason to have just one diary. You can only be in one place at a time. It is dangerous to work a separate diary for your personal life. What about that dental appointment that occurs during business hours? You can't book the appointment at home, because you have left your business diary in the office. Is this really what you want to waste your time doing? You are kidding yourself to think that the two parts of your life are separated so neatly and that there aren't conflicts over your time. In fact you would be able to plan your personal life better if any important engagements are in your work diary. That way, for those important personal commitments you can at least have a 'heads up' on trying to manage that particular day around a personal goal. From the opposite perspective, there is nothing more irritating to your line manager than not being able to tick something off in seconds because someone in the team has left their other diary at home. Some people actually adopt the two-diary tactic or the '*I don't have my diary with me at the moment*' approach in order to get out of making commitments to do particular things. We know who you are!

You should always have your diary available with you wherever you go. It should just be a subconscious part of your ever-present items of extreme importance (keys, mobile, wallet and diary). Every time you go to see a client, every site visit, every meeting. Your diary should be with you all of the time, poised and ready, enabling you to effectively make appropriate arrangements. Just think of what is involved if someone phones you asking for a meeting and you can't schedule it because your diary isn't to hand. First you have to get to your diary, and part of that involves remembering the person's request in the first place. You may need to scribble a note somewhere to remind you. Then you have to call the person back. You might be unlucky and the person you need to speak to isn't available. Then you need to call again, and so on. Yesterday it would have been an answer to a simple question, but today it is out of all proportion. But you have to honour that commitment you made because your diary wasn't to hand. It's just not logical to expose yourself in this way. It wastes time.

What about when you are on holiday? Surely you wouldn't take your diary with you on an overseas trip for a couple of weeks. It's just going to add weight and bulk to your suitcase. You are probably right. But imagine you are away and an important appointment needs to be made for you in your absence. Would you like someone else to take control and decide when and where you will be the day after you come back from holiday, or would you rather have some input and an element of control over how your time is used? For me, it's the latter. Do I take my diary on holiday with me? No, actually I don't. I do, however, take pictures of the following weeks of my diary on my mobile phone just in case something needs to be arranged when I am away. That way there is no bulk, but I have the information should I need it. The reality is that I rarely ever need it. But if something extremely important comes up and I do need to confirm an appointment it's there, it's done and it's far less hassle than trying to remember what's in the diary thousands of miles away.

Many people just think about using their diaries for appointments and meetings. But you can use your diary for all sorts of things. You can block off time in your diary to do tasks, to catch up on things, to make that call back to the person you promised you would speak to by the end of the week. Something I have to do quite regularly is block off time in order to prepare for training courses we are delivering. My paper diary gets blocked off for a day and our team manager is told not to schedule anything in for me during that period. Used this way, the diary helps me to be more effective, by ensuring that I have the time set aside to complete a task.

When you put something in your diary you are making a commitment to yourself, and that commitment can involve more than meetings and appointments. I find it very effective to put certain goals in my diary, be it personal or business-related, because when I look at what's on for that day I always ensure that everything in my diary is achieved. If it's in the diary it will be done.

A diary can also be used as a date- and time-structured method of having your 'to do' list managed effectively. Rather than having one long list on your desk, you break the list down into what you want to achieve over the days and weeks ahead. One less bit of paper (because you don't need the list), and everything in its proper place in the diary, making things more achievable along with a stronger commitment from yourself to get it all done. Then everything else that gets added to your workload gets added to your diary. By adopting this approach you can also schedule in weekly, monthly and annual tasks. These are all items on your 'to do' list, they are just being slotted in for a particular day. If something needs to move, then you move it to the next most convenient time or day.

Now let's go one step further. You have a project on the go that might involve numerous important deadlines ahead of the final deadline for a report to be with the client. You have a plan and a schedule developed in order to achieve this. It should all be documented in the client file. What you should do now is document the key milestones in your diary. This gives you added focus, and is also a buffer should something go wrong. For example, you suddenly take ill one week and your boss asks you over the phone, *'Is there anything you have scheduled this week that needs attended to in your absence?'* Easy: your diary will give you the prompts you need, because in the diary, appropriately dated, are the date you are anticipating the local records centre to get back to you by, the date of the field visits, the date for your proofread to be ready for your manager, the date the report is due to the client, and so on. Not only that, you can go one step further again and front-load your required deliverables by putting early reminders in there. For example, let's say the draft report is due by Friday the 15th, you should add a note in your diary for Monday the 11th, reminding you the draft is due by Friday. There is no point remembering this on the evening of Thursday the 14th. So take these big tasks that you are responsible for and break them down into smaller stepping-stone goals, and get these stepping stones in the diary. Yes, you are going to need something a little bigger than the slim diary that fits into your pocket, but if you do this well you will be so organised, and you will rarely, if ever, forget anything.

TIME MANAGEMENT

The most precious commodity we all have is time. You can't buy back time. No amount of money can replace what you have done to this point in your life with the time you had available. Conversely, how you choose to use your next minute, hour, day or week is within your control. You may say that this isn't entirely correct. You have a job and your boss tells you what needs to be done. True, but technically speaking you can choose at any point to leave. Ultimately, therefore, you are in control. But for goodness' sake don't resign just yet. There may be other options that will make life less stressful.

Once a moment, an hour, a day has passed you are not ever getting it back. So maximising the outputs from the time you have available is crucial. Too many of us, in all aspects of our lives (not just professionally), seem to take time for granted (I'll do it tomorrow; there will be plenty of time later). Yet, how many times do some of us need to burn our fingers in the hot ashes of *'if only I had started this sooner'* before we get the message, *'I am an idiot for doing this to myself'*?

It may not be possible to buy back time, but is it possible to create time? Well, in a way it is. By being really organised you can reduce the amount of time wasted and then use the time you have created to best effect. It's good to think about it in the terms of your behaviours. Figure 5.2 demonstrates that you have a given amount of time to achieve something, and if you have time-wasting behaviours then you are destroying the time you have available. This means that you are creating more pressure for yourself. Conversely, if you adopt time-saving behaviours, you create more time, which in turn should mean less stress and of course a better result all round.

Table 5.2 shows some commonly occurring time-saving and time-wasting behaviours. Try going through these lists and ticking off the things that you do. Anyone would be doing exceptionally well, or totally deluding themselves, if they did not have something in the time-wasting column. But if you can change just one or two of these negative time-wasting behaviours you will have a positive impact

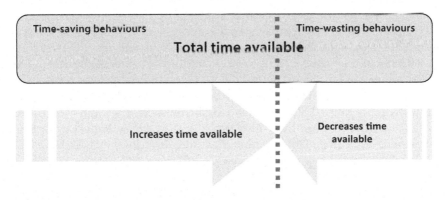

Figure 5.2 The conflict between time-saving and time-wasting behaviours

Table 5.2 Time-saving and time-wasting behaviours within the working environment

Time-saving behaviours	Time-wasting behaviours
Get in early/stay late	Come in late/leave early
Check diary the night before	Don't check diary regularly
Put reminders/prompts in diary	Don't use diary to its full potential
Have a good system for filing/putting stuff away	Lose things
Tidy desk	Have an untidy desk
Tidy files	Messy/disorganised files
Plan ahead	No forward thinking
Set goals/mini-deadlines for each day	No self-imposed goal setting
Adopt a positive mental attitude	Complain and moan a lot
Be punctual	Often running behind schedule
Respect own time and that of others	Disrespect other people's time
Ask for guidance and feedback	Don't ask for and never be open to feedback
Take short lunches/breaks	Take extended lunches/breaks
Use dead time effectively. Think ahead about what goes with you so that things can be done during such periods	Waste dead-time opportunities
	Procrastinate
	Be easily distracted
Make phone calls and check for messages when out of office so that you are more immediately productive on other things when you get back in	Stall rather than ask
	Jump into other people's conversations
	Listen in to other people's conversations
	Make long or regular personal phone calls

upon the time available to get through your day. You will be creating time and therefore becoming more effective than you currently are.

Someone said to me the other day, *'Neil, I know I need to get organised, but at this precise point I just don't have the time to do it.'* I challenged that thinking with, *'If things are really that bad, now is precisely the time to do it.'* You may feel that taking the time to draw up a plan of action, for example, doesn't get you any further forward. But it does. It reduces the potential for wasted time and as such will very often create some of the additional time you so badly need. In addition, it gives you the components required to get the job done. Typically, people who are displaying disorganised behaviours are poor at planning, and it is these same people, when they finally swallow the get-organised pill, who stand to gain the most. They don't see it, possibly because it's just never been their approach. OK, I said they are poor at planning. Perhaps that's not entirely fair. Planning isn't difficult. It doesn't take a huge amount of intellect or effort. I suppose it would be better to say that they don't bother planning as they have yet to appreciate the benefits of doing so.

Surely no one enjoys wasting time, losing things, working in an untidy environment. Some people say they are happy as they are in this respect. They say that they have managed so far being this way. They say that they are too long in the tooth to change. I would argue that there are two conflicts. On the one hand, you manage your time, and on the other, time manages you. Successful people manage their time. For those of you who are already successful despite being disorganised, I would challenge you by saying, *'Oh my word, just think how much more successful you could be if you were organised!'* You are underachieving, and doing yourself a disservice.

You can choose to use your time productively, with everything planned out to the minute as to how you can achieve that result (no excuses). Or you can choose to just let things happen in a far more casual way, and if you get to Thursday and that report due on Friday is nowhere near finished, you can always work through the night to get it done. Just think about the logic behind this latter approach. If you had the choice of working a couple of hours each evening for three nights or working one evening into the small hours, what would be your preference? Personally, for jobs that are going to take up a lot of time I am a huge fan of getting started as soon as possible, and doing smaller achievable chunks in order to arrive at the result. This gives you more time to think about what you are doing and to take on board any feedback that makes it even better. It is a classic problem-solving method. Break the problem down into smaller components and knock them off one by one.

Procrastination is the thief of time
Edward Young (1683–1765, English poet)

OK, I have been putting off writing about this for some time now. I know that I have needed to do it, but it was going to be difficult for me, and it wasn't really that urgent (at the time), and I had other things that were more important during each of the past 45 days when I could have at least made a start on this section, and I was ready to do it the day before I headed off on holiday, but something else came up that I knew I could get cleared off my desk in no time. And breathe! So, here I am now, typing away and I'm not really looking forward to the next couple of hours. I knew it was going to be difficult, but now I have no choice, it needs to be done and by heck it maybe wasn't that urgent a few weeks ago but it's certainly urgent now. A blend of urgent and complex, a recipe for disaster! Hold on ... I'll be back in a few minutes, something else has come up. Honestly, it will just take a minute ...

Can you relate? I am sure that most people can. I certainly can, and the number of times I have put difficult tasks to one side in favour of other simpler, quicker, nicer things to do is countless (especially in my early days working in an office environment). However, if you are one of those people who regularly encounter this, have you ever noticed how, when you actually come to tackling the dreaded task of doom, it's rarely anywhere near as bad as you imagined it to be? Now there is a very good reason for this. What tends to happen is that we see the thing that needs attending to, and for whatever reason we put it off. It may be because it's complicated, or at the other end of the spectrum it's just plain tedious and will take a lot of effort. Usually, either way, at the time it raises its head initially it's probably not that time-critical (your manager says they want it in a couple of weeks). It's probably not that important compared to other things on your desk. But you know it's there, and worse still your manager or a customer knows it's there. There is an expectation from elsewhere that it will get attended to. You keep putting it off, and all the time you are doing this, subconsciously the task is getting even bigger or more complicated (or both) in your head. Meanwhile, tick, tock, Father Time is creeping slowly (is it possible to creep quickly?) towards the moment when it can't be put off any more ...

... And then finally you take a deep breath, make a cup of your favoured brew and just get on and do it. It may be even less pleasant now, as you may have someone breathing down your neck, applying pressure. Hours pass and the task is completed and signed off. In your head you say to yourself, '*Goodness that was nowhere near as bad as I thought it was going to be. I don't know why I just didn't do it weeks ago, and if only I had I wouldn't have had the additional, unhelpful time pressures being applied.*'

There you have it. In short, if something needs doing, just get on and do it, and if it is one of those rare scenarios when it actually is as bad as you think it is going to be, at least it's done. At least you don't need to think about doing it for weeks and getting stressed out about it. At least your manager will be pleased to see that you have tackled the situation full on and the box can be ticked, and everyone can move on to something better.

What I tend to do (and in fact I have done this a few times while writing this book) is to give myself a reward for completing the task (or a stage of the task). Now, to stress here, it's a reward. So quite genuinely, at this precise moment, I am typing this in a hotel somewhere in deepest England, and before I went to bed last night I planned today. The plan was to get up, have breakfast and check emails. Then write for four hours, and then, once I have achieved that, I will go (haven't been yet – still typing) and spend an hour in the gym and then have a swim. For some that may not sound like a reward, but for me it's heaven. Now I could have turned the day on its head and done the gym/swim thing first, but that just wouldn't make sense in my world, as I would then spend all that nice time wondering how much book I would actually go on to write. Would I write any, or would other work-related stuff get in the way? When I get to the end of the day I don't want to be reflecting upon what might have been. I don't want to be making excuses. The reward has to come after the effort has produced the desired result. Otherwise, I wouldn't feel I had earned it.

Furthermore, and probably even more importantly, as it can have a direct impact on customer relationships, procrastination does not relate just to written tasks. It very much relates to delivering bad news either by phone or face to face. You can of course deliver bad news by email or letter, but generally speaking that's not a professional approach. If something has gone wrong or has not gone to plan, and you need to break the bad news to someone, the most professional and credible – and the kindest – approach is to lift that phone and make them aware of the situation. And do it NOW: it is the most important thing you are likely to have on your desk. You really don't want to phone the developer to tell them that on the final bat survey, carried out last night, you have found a bat roost and it means they are not going to be able to knock that building down next week after all. There will be delays, and costs, and mitigation, and perhaps even compensation. They are not going to be happy. Oh my word, they may even shout at you. They may ask to speak to your boss or they may never use you again. Really, is all of that the worst that can happen? Well here is the news – it isn't. Things could be even worse than that – but I'll get to that in a minute.

Here is the reality. When you lift that phone and tackle it head on (even if you fumble nervously through the message you are delivering), it is never as bad as you think it is going to be. And if it really is that 'once in a lifetime' scenario when it is actually as bad as you expected or worse, then all I would say is thank goodness you made the call. That was the one call that really needed to be made, and the chances are that the person you are calling is justifiably angry with you or your business. In that case, it is so important, for so many reasons (e.g. feedback, evolution of process, self-development and experience) that you hear it full on.

But a few moments ago I said things could be even worse. Here goes, this is what would make it even worse: procrastinating for hours or days before you have the conversation. Now you have a bigger problem, because not only are you giving them the bad news, you are also having to explain why you have been so inefficient and apparently unconcerned and unappreciative (this is the perception of the recipient) of their world that it has taken you two days to think about making them aware of the situation. In that set of circumstances you may very well have an angry or disappointed customer, not because of the initial message, but because of your perceived lack of urgency about the matter, or lack of respect for the relationship. And who created this situation? You did, by not dealing with it as quickly as you could have. Who is to blame that the customer is angry and you are now having to defend the indefensible? You are. Who can help ensure that this never happens? You can.

So, in summary, procrastination is inefficient at best and deadly to business relationships and your credibility at worst. Spend some time thinking about this (not too long, though) and grab whatever it is full on and fix it today. No matter what, no excuses. When you eventually hit the pillow tonight (or early doors tomorrow morning), remember how you are now going to reward yourself – and sleep well!

Phew! I am glad that's finally out of the way. Wasn't anywhere near as bad as I thought it would be. Time to hit the gym.

Deadline management

We all dread a looming deadline, and in the world of ecology they are an almost ever-present fact of life. You have to get that job done on time, and there may very well be repercussions if you fail to deliver. And of course it's not just about being on time, it's also about delivering something that is 'fit for purpose'. We can all do a bad job fast.

I was recently reminded of an expression which I first heard many years ago: *'If you want to do something, you'll find a way. If you don't, you'll find an excuse'* (James Rohn, 1930–2009, entrepreneur, author and motivational speaker). This can relate to countless scenarios, and I am going to use it here in relation to tackling complex tasks or things that don't initially appear possible to achieve. Let's say you are given a task to do by your line manager (e.g. Phase 1 survey, with a reporting deadline a week from today). You might feel that the deadline is tight, bearing in mind everything else you have on your desk. However, if you *really* want to do it and if you set about putting together a plan for completion of all of your tasks you will at least be giving yourself the best chance of success. Alternatively, if you aren't that motivated, or if you start from *'it's just not possible to achieve'*, then at the end of the day you are potentially only preparing for the delivery of an excuse.

So how do you go about developing a plan for success? Table 5.3 gives an example, taking the latter stages of the report already referred to through its last 24 hours. The important thing to remember when doing this is to start with the end result in terms of time, and then work backwards to where you need to be and what needs to be done at various points leading up to the deadline, slotting in the actions and mini-deadlines as appropriate. My partner is sometimes surprised by how much I can achieve in a short period of time, and this is precisely the technique I use when multiple things are needed by a certain point in the day. By working backwards you actually see, more often than not, that it can be done. The only question that then remains is: Will you just get on and do it, or will you make an excuse?

It's all about demonstrating to yourself, and sometimes to others, that it is possible. Yes, something else may need to be sacrificed in the process, but that is where

Table 5.3 Planning and setting targets for urgent tasks (example: report deadline).

Technique	Deadline	Required actions
List things starting with the end point in mind and working backwards from then until now	Tomorrow, 10.30 hrs	Get it done
		No excuses
	Tomorrow, no later than 10.00 hrs	Submit to manager for proofread
	Tomorrow, < 09.00 hrs	Print off and proofread myself
		Make any required changes
	Today, < 17.00 hrs	Complete report
	(meeting at 15.00 hrs)	(Remember 20-minute meeting with senior ecologist for their thoughts)
	Today, < 13.00 hrs	Maps back from GIS technician
		Check and insert into report
	Today, < 09.15 hrs	Remind GIS technician I need maps by 13.00 hrs
		Remind senior ecologist re case meeting this afternoon
		Start completing 'results' and then 'discussion' sections

redefining your priorities comes into play. Quite often it may not even be your call. Your manager may decide that it is acceptable to allow something else to slip, in order to get this new and more important piece of work completed. So never assume that what initially appears impossible can't be achieved. Find a way. Ask yourself *'What needs to happen in order to make it possible?'* It may involve working late, it may involve something on your desk being passed to a colleague, or it may involve phoning other clients and establishing if any of their deadlines can be shifted back a bit. One thing is sure, however, and that is that strutting about the office moaning about it and then going away to the pub for an hour in a huff isn't going to achieve anything. Nothing is going to change apart from your ability to deliver an excuse. At the end of the day you may have the *'I told you so'* satisfaction, but no one who matters is going to think that you really pulled out all the stops. They are just going to feel that your negativity was a contributory factor to the failure.

We are a service-sector business. The client requires the service your company has committed to. Make it happen. Thank goodness not every week is like this. But for the next couple of days it needs to be a concerted effort. Get in early, work late. Organise yourself and plan it out. You will surprise yourself as to what is possible if you really want to make it happen. This is part of the job satisfaction that makes your career right for you. This is your manager now thinking so many positive things about your character and your commitment to helping the business succeed.

Ask, don't stall

'Why didn't you just ask me where it was, rather than spending an hour looking for that report template?' The amount of time that gets wasted and the hours that get eaten into because people are so determined to be self-sufficient and not show what they imagine will be perceived as weakness, is beyond me. In our business we get very nervous about people who don't ask questions. It has to be stressed that to admit that you are struggling to understand something or that you don't know where to find some piece of information is not a weakness; it's a strength. It demonstrates that you know that time is valuable and that the business needs you to get things moving as quickly as possible.

This doesn't purely relate to the initial stages of a task. When you are further into a task, it's good to check with someone that it is being done as envisaged. If you are off on the wrong track it is far better to find out a week before, rather than an hour before something is due. People who sit and say that everything is 'fine' (never a good word!) and they don't need help are the ones to be really cautious about. Especially if they have a track record of failing to deliver. They may actually need to be shown that many heads are better than one, that they are part of a team, and that if they are looking for a solution the chances are that someone else in the business has encountered something similar in the past.

So if you find yourself stalled at a particular point, ask for advice immediately. If you find that something is taking a long time to complete, ask if there is another approach that can be taken. Don't just sit there achieving nothing or adding to the problem. Think about it. Asking someone where the template is may take up a couple of minutes of their time. Going through lots of folders looking for it in the hope that the one you have chosen is the best fit for your report on this occasion may take an hour. What is the logical thing to do? What is the most cost-effective use of everyone's time? What would the owner of the business want you to do? Ask, don't stall. The one thing to bear in mind, even more so for more detailed discussions, is that when approaching someone else in the business for assistance, it is a good idea first of all to ask when would be a good time to speak to them. Apart from being good manners, this means you are more likely to have their best attention when they are assisting you.

While we are on this subject, another similar behaviour is seen in those people who assume that they already know everything there is to know about a subject or a job. This is a very dangerous frame of mind to be in, especially during the early years as an ecologist. If you are one of those people, you would be far better advised to assume that there is lots still to learn. There will be people around you who have considerable experience relating to the type of projects they are involved with. To assume that these people cannot help or give you a good steer is to deny yourself so much, as well as potentially exposing you to risk. People that have been in the sector all of their lives are learning all of the time. It never stops. But it certainly starts, and that starting point is to consciously seek the opinions and guidance of those around you.

Time sheets

Within many ecological consultancies time sheets exist whereby those working within the business can accurately record the time that is being taken on specific tasks or in relation to specific projects. These sheets allow managers to see how long it really takes to carry out a piece of work, which goes on to be reflected in how much the client will be invoiced. How long your line manager thinks a piece of work will take, and how long it actually takes, can very often be completely different. The information you record must therefore be accurate for it to be of any use when determining in the future how to cost jobs or to establish which activities are more profitable. You must therefore factor in the time to fill these out, and it must be done correctly. If you massage the truth about how long something has taken, you are denying your employer income. In addition, you are doing yourself a huge disservice. Let's say that you put down that a task has taken you five hours when in fact it has taken seven. Guess how long you are going to be expected to do the same task for another client next time. Guess who is going to have to explain why it wasn't completed as quickly. You must be credible and honest with your employer about what you have been doing with their time.

One added benefit of this system is that it also allows you, personally, to see where you have spent your time. By doing some self-analysis in this respect you may very well find areas that can be improved upon.

MANAGE YOUR PRIORITIES

One of the essential skills in managing time lies in effectively prioritising what you have to do. It is so important to do things in the right order, which means doing what really needs to be done first, next.

A good way to approach prioritisation is to use the matrix shown in Figure 5.3, which was created initially by Stephen R Covey (1989). In effect you should insert the items from your 'to do' list into the appropriate boxes within the matrix. Having done this, the things you should be working on next should become considerably more apparent. In short, you do what's both important and urgent before you start on anything else. As importantly, tasks that are either not important or not urgent will become more apparent and can be scheduled into your workload accordingly. Overall, you should now be better informed as to what to do and the order in which things need to be tackled. In doing all of this, also bear in mind that the tasks being considered could in fact all relate to one larger, multifaceted project.

Having allocated the tasks to the different boxes, you may have *opportunities* and *conflicts*. Opportunities may occur whereby you are able to delegate tasks, or components of tasks, to others in order to speed things along or free up your time to focus on other areas. Conflicts may arise when different tasks each appear to have the same *'do it now'* status. In such circumstances you just need to crack on and get all of the urgent/important tasks completed. But if there is a true conflict, and doing one task is going to impact upon your ability to do another within a prescribed timescale, then you should immediately highlight the problem and ask your line manager for guidance as to which one takes precedence.

	Important	Not important
Urgent	1 **HIGHEST PRIORITY** Do it now	3 **LOWER PRIORITY** Question: If it's not important why is it urgent?
Not urgent	2 **MEDIUM PRIORITY** Make progress on this Schedule in time Eventually it will be urgent	4 **NOT A PRIORITY** Do it later if at all Question: Does it need doing at all?

Figure 5.3 Time management (prioritising) matrix (adapted from Covey 1989)

In regards to the matrix there are a number of other points that should be considered. Firstly, look at those items which appear to be both 'not urgent' and 'not important'. You should really challenge yourself and others on this. If it's 'not urgent' and it's 'not important', does it really make any difference if it gets done at all? Are you going to get into trouble if you don't do it? If the answer to that question is 'no', then you should really be querying why it's there in the first place. Sometimes in situations such as this it turns out that the task is more important than you initially realised, it's merely that you haven't been given all of the background behind it, or you have misinterpreted what you have been asked to do. In these scenarios, thank goodness you asked. The point is that you shouldn't really decide that you are not going to do something you have been tasked with without first having a conversation about it with your line manager. Secondly, the 'urgent' but 'not important' stuff. My first question would be: *'If it's not important, then why is it urgent?'* So please ask that question. The points mentioned in the previous paragraph still apply. There could be a reason that you are not currently aware of. Finally, the 'not urgent' but 'important' category. The one thing that is certain is that it may not currently be 'urgent', but if it's truly 'important' and you haven't done it then one day it's going to become 'urgent'. It always makes sense to keep an eye on these items and take every opportunity to make progress towards their completed state.

When tasks are complicated, or time-critical, it is really important to be moving things forward as quickly as possible. This can happen when you have a task that involves other people's input or advice. Figure 5.4 provides an approach to use when others are involved with you successfully hitting your deadline.

It is always an idea, irrespective of whether you are going to be doing anything right now (e.g. in a not urgent/important situation), to quickly get things moving along with the other people you are going to rely upon. In doing so you can at least be confident that things are moving forward. The important thing is that you

Task characteristics		
Lengthy or complex task **Relies on input from others**	**Quick and simple task** **Does not rely on others**	Deadline
Action	Action	
Start right now	Leave to nearer deadline	
Create plan and set targets	Lump together	
Delegate now	Try and integrate these tasks into 'dead-space time' created while working on separate, unrelated tasks	
	Do earlier if no impact on lengthy or complex tasks	

Figure 5.4 Prioritising urgent and important

quickly pass those components on with clear instructions, and of course informing the other person of the deadline by which you need the response (see Appendix 2, *Effective allocation of tasks*). This approach would apply whether it was a colleague that you needed to complete something, someone in another department, or indeed a request to an external organisation. For example, you might be doing a report within which there will be a desk study. The desk study requires data from the local records centre. Day one, send a quick email to the records centre asking for their records. You may not be planning on starting the report for a few weeks, but there is no logic in not getting the other people you will be relying on to move forward with their contributions.

The creation of time

I suggested earlier in this chapter that it was possible to 'create time', and what I have just described is another example of the same sort of idea, where multiple tasks are being dealt with simultaneously. This is all about getting things moving concurrently for tasks you are responsible for while you are engaged with other aspects of the same task, or with different work altogether. In other words, things are all progressing in the right direction simultaneously.

This can be achieved either by having other people or systems working for you while you are doing something else, or through managing your own time in order to have two or more things within your direct control (without the involvement of others) occurring at the same time. For the latter, you need to put yourself in a mind-set that you can actually attend to two or more things simultaneously, as opposed to doing them one after the other. When opportunities to do this occur, usually one, or both, or more of the tasks are fairly straightforward, but potentially time-consuming. Opportunities to do this present themselves all the time, it's just a matter of being alert to the concept and aware of all the things that you could be doing at any one particular point in time.

To demonstrate, let's look at a straightforward, everyday example (Figure 5.5). You have a large document that needs photocopying and you also have to make a quick phone call to a client. Grab your phone and go to the photocopier. Stick the document in the automated copy feeder, walk away and make the phone call. You then find that you are put on hold waiting to get transferred to the right department. While you are waiting to be connected, walk back to your desk, call up your junk mail inbox and start deleting.

Now I am aware that I am taking a fairly straightforward scenario to demonstrate the point, but I actually know people who would go to the copier and wait for that to finish, then go back to their desk and make the call, and while waiting on hold it just wouldn't occur to them to be using that dead time to get on with something else.

In the example given in Figure 5.5, what would have happened if you changed the order of the tasks and made the phone call first, then did the photocopying, and then later in the day remembered it was about time you tidied up your junk mail

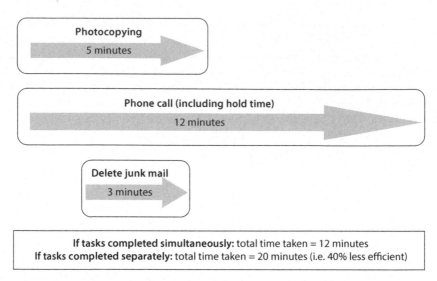

Figure 5.5 Multiple task delivery: the benefits of doing more than one thing at a time

inbox? In effect by doing things in a different order you would have denied yourself the opportunity to make savings in time.

Taking this approach may only appear to save you minutes here and minutes there, but all these minutes will add up. Apart from that, you will be getting all of these distraction tasks ticked off far more quickly, leaving you with more time and a clearer, more focused mind for the more complex stuff.

Making good use of what initially may appear to be unproductive time is also extremely beneficial. An example of this would be to put all of your background reading into a file on your desk (less cluttered and all in one place). Think of occasions when you can do some serious damage to what's in that file. For me it would be occasions when I am waiting for a train, in an airport lounge, sitting on a ferry or being forced to take a long lunch break while waiting in my car for a contractor to come back to a development site. These, and many more, are all examples of situations where you could end up doing nothing of any real value for a reasonably long time. By taking advantage of that dead time today, you are going to reap the benefits of this, days or weeks later. It's just a matter of having your stuff already organised (everything in its right place) so that you can immediately lift something to take with you as you leave the office, in anticipation that there may be time to do it.

Using this dead time to catch up on emails (if you can access these via your mobile), to phone the office to check if anyone is looking for you, or to make those calls that you were given to follow up on yesterday are other examples of how you can use your time more effectively.

It's a false economy to ignore what might be going on back at the office. If a note to phone someone has been left on your desk in your absence, for example, then it's coming your way and you are going to have to deal with it. It is far better that you take control of when you deal with it, and you do it when there is nothing else that can be done, as opposed to you walking into your office tomorrow morning all ready to get started straight away on something, only to find you have had an immediate distraction forced upon you.

Are you the type of person who would quite happily, during work time, sit in that airport lounge catching up with your social media? Fine, be that person. But when you are back in the office metaphorically throwing furniture about because you have too much to do, stop and think why you are in this position. You have had the benefit of the easy day travelling to a meeting. That was your choice. Now it's 'pay back' time. Alternatively, use the dead time to your advantage, and tomorrow will be that little bit easier.

ORGANISATIONAL EXCELLENCE

Once you are truly good at 'being organised' you are on the way to achieving 'organisational excellence', which could be defined as follows:

> The ability to effectively take on new responsibilities and challenges, without losing the ability to continue at the same effective level of performance those things you are already responsible for doing.

A big part of achieving this level of excellence is not just about developing the skills and techniques, but also about knowing when you have reached your limit of effectiveness. It is as much about saying 'no more' as it is about saying 'yes, I can do that'. Unfortunately during your paid job you don't have too many opportunities to say 'no' to your line manager. However, you can control other aspects of your life, and you can decide when or when not to volunteer for opportunities at work. Accordingly, I am going to conclude this chapter by asking two questions:

> Is it always a good idea to volunteer to take on additional responsibilities at work?

> Is it always a good idea to accept invitations or volunteer to take on new roles beyond the parameters of your main paid employment?

The answer to these questions depends entirely on the amount of spare time that you *realistically* have available, and on your ability to remain organised and in control of the areas that you are already being tasked to deliver upon, presumably to a high level.

For those people who just don't have the time available and/or who aren't organised to a high level, there is huge potential for them to place themselves, unwittingly, in a position where they have too many responsibilities or roles, or simply too much going on in their lives, and they don't have the time to do justice to everything. As a result they risk watering down their effectiveness in each area of responsibility, and hence not being successful at anything.

You need to recognise that there are only so many hours in a day, and for many of us taking on additional roles means that less time is spent on other things. Can you afford to spend less time elsewhere? Perhaps you can, and if so the extra role may work very well for you. Just be realistic at the outset. Good advice is to take things on selectively, grow into them until you are at a high level of competence, and only then take on something new. At this point you will hopefully have the ability and the time to give the new responsibility your best shot, without destroying what you are already achieving elsewhere.

I have seen too many people spread themselves too thinly across too many responsibilities, and then find that they fall short of meeting anything close to their true potential. I have done it myself. I have bitten off too much and, somewhere, something else has suffered. Today, I don't take on anything additional unless I can truly see how it fits into my work–life balance, and that it won't have a negative impact elsewhere. It is often so pleasing to be invited to do new and exciting things, that we all can be flattered by the well-placed sentiments of the requester and just say 'yes'. I think the key is only to say 'yes' selectively, and only to do things that you can truly add value to.

Chapter 6
MEETINGS, MEETINGS, MORE MEETINGS

Alone we can do so little. Together we can do so much.
Helen Keller (1880–1968)
American author and social activist

After being in the field for what seems like weeks, you finally get a full day in the office and you have it all planned out. Everything that needs to be done that day. Everything that has been filed in your *'I'll do that when I am next in the office'* folder. Yes, today is going to be extremely productive. It has to be, because you are back in the field tomorrow gathering even more data that will need to get typed up and reported upon. You get in early, and you get off to a brilliant start (no one else has arrived yet!). Then your line manager comes in. *'Brilliant, you're here. I need you to come into a meeting with myself and a client later this morning. It will only take an hour or so, but you will need to read over this file before we kick off.'* Your heart sinks. The day is already evaporating before your very eyes. Hands up if you can relate to this. My hand is also firmly in the air. Sadly not only from the employee's perspective, because I am probably, on occasions, one of those managers.

Meetings – they just get in the way of everything else. And why do we need to have so many? Well unfortunately (if we remain in negative mode for a short while longer) they are sometimes very necessary. For example (let's now get positive): in order to get things done efficiently; in order to ensure everyone gets the same message at the same time; in order to collectively share thoughts and exchange ideas. That being the case, why do they have such a bad reputation?

When meetings cause negativity it is usually when they take place for meeting's sake. The wrong people are invited. The right people don't turn up. The subject matter could just as easily have been delivered by email or over the phone. When these sorts of things happen it really is quite irritating. I remember once I had to travel all the way from Glasgow to Birmingham to hear a presenter deliver a PowerPoint presentation which he, pretty much, read word for word from the slides. Afterwards, when he was taking questions, he had no further information to give us over and above what was on the slides already. Approximately 20 people attended that day, each travelling from a different part of the UK (including someone who flew over from Belfast). What a complete waste of time and money. Yes, we all got the opportunity to catch up with colleagues, and for some carefully woven networking, but it was not the best use of our time or the company's money.

Having said all of that, I can also recall countless meetings that were precisely what was required, and by a long shot were the best way to achieve the objectives set. They were well conceived, well managed and instrumental at key moments to the success of projects. What made these different? Why were they good meetings and an excellent use of time?

IS A MEETING REALLY THE BEST SOLUTION?

For those of us who are fortunate enough to be in a position whereby we can use our status to call meetings, there always needs to be that question in the back of our heads. Why does it have to be a meeting? Is there not another way in which the desired result can be achieved as effectively without tying up so many people's time? Sometimes it's too easy to say 'Let's get everyone together in the same room and talk about this', when the smarter thing to do is to ask 'What is our objective and how best do we achieve what's required?' There are usually other options available which may very well be quicker, cost less money, and still achieve the desired result. For example: an email; a video conference; a conference telephone call; speaking to people on a 121 basis. It all stems from what the ultimate goal is, and the decision should be based on weighing up the pros and cons of different approaches and settings.

Is it a situation where you have a message to give to your whole team about something that is happening, but which doesn't really directly impact upon them? If so, perhaps an email will convey the message quite easily. Alternatively, is there some big new idea that is going to directly impact upon your team and you genuinely want them to have a collective discussion, so that they can react to each others' ideas and help formulate the way ahead? If that is the case, then yes, a meeting is going to probably be the best setting within which to achieve this. Provided of course it is pitched appropriately and organised in the right way.

ORGANISING THE MEETING

Having established that a meeting is the best format for the goal to be achieved, then the next question should be, what is the best way to invite people? The meeting may very well be required urgently, and it may involve people who are all in the same location as yourself. In that case, bouncing into the room and shouting, 'Everyone stop what you are doing. We all need to go to the conference room for a meeting. Now!', may be effective, at least to begin with. If, however, you begin to develop a reputation for holding meetings like this, for objectives that aren't really that important, then very quickly people are going to despise you, the conference room and all that your meetings entail. Therefore, in the longer term perhaps not that effective after all. The best meetings are usually well thought out and carefully planned a reasonable time in advance. Accordingly, an email inviting attendees would most often be the best method to invite a group of people. This email should be carefully crafted to include all of the details that an attendee would need to know. It is also very important, in choosing who should attend, to ensure that those who are invited are going to properly contribute towards the objectives set.

The email should state clearly what the purpose of the meeting is, where it is being held, what time it starts and how long it is expected to last. It would also normally be wise to make everyone aware who else is attending. Other things to consider include instructions as to what people attending the meeting need to consider or prepare in advance. Depending on the objective, it might be appropriate to invite attendees to put forward suggestions for any additional items that could be catered for on the agenda. Confirmation of attendance would also be normal, especially if there are people coming from other locations or other departments, or from outside of the business. In these situations those being asked to travel should be advised as to whether car parking is available and if so, where.

The timing of the event needs to be governed by how important it is to have it sooner rather than later. The main thing to think about here is to avoid times that are really inconvenient for proposed attendees. For example, if it is a one-hour meeting that doesn't need to be held at 09.00 hrs on a Monday morning then avoid this time if at all possible. If the meeting is that important, then it's beneficial that those attending are in the right frame of mind for contributing productively. First thing Monday morning, most people are trying to get their week off to a flying start. Last thing on a Friday, most people are trying to get things closed off ahead of the weekend. So think about who you are inviting and why, and if certain times of the week are better for having the meeting in order to get the best result.

As well as when, you should also consider where. For some businesses there may not be that many options, and indeed if everyone attending is based in the same office then the where is usually self-evident. What, however, if you are a multi-office business or you have remote workers? Let's assume you have people based in Glasgow, Edinburgh, Aberdeen and Newcastle who need to attend this meeting. Aiming for a 09.00 hrs start in the city centre where you live might be good for you, but how frantic will it be for those travelling from some distance away. If it doesn't need to start at 09.00 hrs, put it back an hour. Better still, perhaps, hold the event at a motorway hotel meeting room on the outskirts of the city where everyone can get to it far more easily, as well as removing parking challenges. It might cost a little bit of money to hire the venue, but add up what the costs of the extra city-centre traffic time would be, the parking costs, the extra business mileage and the time saved after the meeting enabling everyone to get back more quickly to their normal day-to-day responsibilities. Not that expensive after all, if you look at it like this.

Who to invite seems really obvious. However, sometimes we just fall into the same old, same old frame of mind and don't really think this through. To have a truly effective meeting you should really be aiming to ensure that everyone who can add value to the objective is there. Having identified the people who you really need to be there, you should go on to consider if there is anyone beyond that group who may, for other reasons, be interested in attending. Provided their attendance is not going to impact on your ability to achieve the objective then by all means give them the opportunity to decide whether or not they wish to be there. Finally, you must consider anyone whom it may be 'politically' appropriate to include within the list of invitees. For example, you probably would not want to have the owner of a major supplier at a meeting on your premises without at least making the owners of your own business aware of this and giving them the opportunity to be involved. You really have to keep yourself 'risk-free' in such circumstances.

It is essential that a proper agenda is created and distributed ahead of the meeting, and as I have said already it is also of benefit if everyone knows who else may be attending. This can be done separately, or, if only a handful of people are invited, within the agenda documentation. The agenda should clearly state the purpose

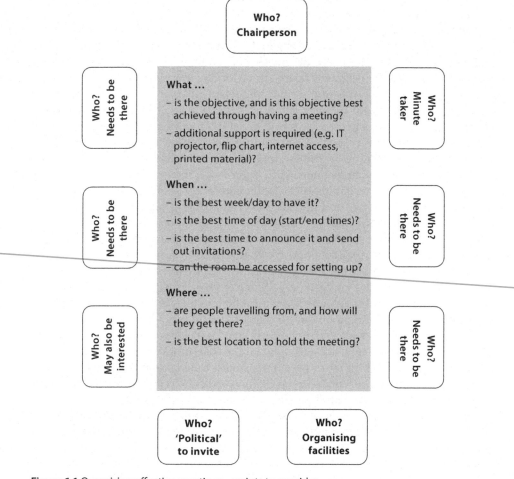

Figure 6.1 Organising effective meetings – points to consider

behind the meeting, including the relevant responsibilities for the event (e.g. chairperson, minute taker). Each item on the agenda should be clear and have a time allocated to cover it. Also bear in mind that shorter meetings (e.g. one hour) with a smaller number of agenda items (e.g. four) tend to be far more productive than cramming in lots of other items and thus making the meeting considerably longer. Even if it's one of those regular team meetings, it is far better that everyone leaves with a clear understanding of what has been achieved and what is going to occur as a result of what has just been discussed. As opposed to them leaving, having discussed 15 items, with everything now watered down so thinly that they barely remember the initial purpose of the event in the first place.

To help focus your mind on the organisational aspects of meetings, Figure 6.1 covers many of the points discussed so far, along with a few others.

ON THE DAY

Before the meeting gets started it makes very good sense to check that the room is set up in a way that is best suited to what you are trying to achieve. Are the seats positioned for best effect? Does the technology work? Are the handouts ready? I am surprised how often people holding meetings just turn up and then mid-flow discover that something quite basic isn't what was expected, resulting in them fumbling through what was supposed to be a powerful tone-setting presentation. This is basic organisational foundations. Get the simple stuff right and the harder stuff will come more easily. If you live in a world where you expect to turn up precisely on time and everything will be perfectly as you expect it to be, then be prepared to be disappointed, and often. For those of us who live in a world where the sky isn't purple with yellow spots, we work hard at getting the simple stuff right. We turn up expecting everything to be wrong and anticipating that we will need time to fix it before everyone else arrives.

There may be different approaches to the room, depending upon who is attending. Is it purely an internal event or are you having guests from another area of the business? Even more so, you should carefully consider the impression created by the room if you are expecting a customer or a supplier to be in attendance. In the latter instance a much higher level is required. This would include paying particular attention to matters such as punctuality, tidiness, refreshments, confidential files being out of view and your colleagues being made aware that an important visitor is going to be in the office that day. It is therefore crucial that someone considers who is attending from beyond the normal locally present staff, and just stops and thinks for a minute about whether anything needs to be different that day. Are there opportunities aside to the meeting that can also be integrated into the visit? Often in this scenario there may be an opportunity to show the VIP around the office and introduce them to other people they may be corresponding with beyond those who are attending the meeting. On such occasions best impressions all round are essential at every level within your team – from the car park all the way through your working areas and into the meeting room.

I will now move on to consider the meeting from the perspectives of those attending, in particular the chairperson, the minute taker, the attendees and those involved with formal contributions of agenda items.

The chairperson

In the chair, you are responsible for the smooth running of the event and ensuring that the overall objective of the meeting is achieved. You will also be the person that everyone else would normally look to in order to maintain focus and ensure that time is under control. If things go off on a tangent or discussions get bogged down in technical details that everyone does not need to hear about, it is your responsibility to move things on to the next logical step of the process.

You need to set clear rules for the meeting. You could liken this to an umpire talking to players ahead of a competitive tennis match. For one-off meetings with groups of people who don't normally come together in this way, just set the tone by saying how you expect the meeting to be conducted. For regular meetings with your colleagues, you can very quickly give a broad understanding as to what you are prepared to tolerate as chairperson and where your boundaries are relative to positive contributions tipping over into protracted distractions. And they shouldn't need to be reminded about this at the start of every meeting. This all sounds very stiff and boring, but it needn't be. There is no reason why people shouldn't engage in humour and light-hearted moments during the discussions, provided of course that the end objective is being achieved. On the other hand, there is no point wasting everyone's time with some hilarious anecdotes, if at the end of the day you all walk out of the room no further forward.

Table 6.1 provides some guidance as to the specific things to consider if you ever find yourself chairing a meeting.

Table 6.1 Chairing meetings – ground rules

Rule	Additional notes
Start on time	The meeting needs to start on time. Attendees strolling in late is not professional and is disrespectful of your position as chair, and also of their colleagues' time.
	Start the meeting on time and people will quickly learn not to be late.
Diary, paper and pen	Everyone attending should have note paper, a pen and their diary with them.
Introductions	Does everyone in the room know each other, and do they understand each other's roles, responsibilities and the reason they have been asked to attend?
	Introductions need to be quick. If the meeting is being attended by lots of people it may not be practical to do this.

Table 6.1 cont.

Rule	Additional notes
Rules	Highlight any rules that apply during the meeting (e.g. confidentiality).
Technology	No technology allowed in the room (e.g. mobile phones, tablets) except that which needs to be used during the meeting (e.g. laptop and projector).
Objective	Remind everyone why they are there.
Agenda	Review the agenda against the context of what the objective is.
	Check that no one has anything else related that should be considered.
	Stick to the agenda and don't allow 'off piste' subjects to encroach upon the time that needs to be dedicated to the objective.
Timings	Ensure the most important agenda items are tackled first. This means that if the meeting runs out of time you have covered the most important stuff.
	Don't spend huge amounts of time on any particular aspect. Keep the momentum going. If things begin to get bogged down then allocate an action point for it to be dealt with at a separate time.
One person at a time	There should only ever be one person talking at a time.
	Don't tolerate people speaking over each other.
	Don't allow whispered side talking to occur.
Participation	Ensure that you allow everyone to have a voice.
	For those attendees who are more introverted, go out of your way on occasion to specifically ask if they have any ideas that should also be considered.
Any other business (AOB)	If subjects arise that are beyond the scope of the meeting then 'park' them to be dealt with at another time, allocate them to someone, to be explored further and reported upon later, or tackle them under AOB at the end of the meeting (but only if there is time, without running over).
	Do not be cornered into adding someone else's agenda into the meeting time. If it's important enough then it should be given its appropriate place and emphasis using whatever setting is best for it to be properly discussed.
	AOB is usually good for 'quick hit' decisions or announcements. It is not a good approach for covering matters that would warrant being an agenda item on their own on a different day.
End on time	End the meeting at the allotted time, unless no other sensible course of action is appropriate.
	Respect that people in the room will have planned their day on the basis that the meeting is going to conclude at a certain time.
	If you have chaired the meeting effectively, you should have managed to get the important items on the agenda attended to, with anything else being safely 'parked' for another day.

There are a couple of other things you should pay particular attention to once the meeting gets under way. Firstly, have you appointed someone to take minutes? Having done so, explain to them what level of detail is required and also give clear instruction as to when you would expect to see the minutes distributed, initially in draft form and then in their final form. It would be preferable that the minute taker is not someone who is likely to be heavily involved in the discussions, because taking the minutes will impact upon their ability to contribute. It makes sense, therefore, that the person is identified and briefed ahead of the meeting, if for no other reason than it will save everyone's time.

Secondly, you need to ensure that any action points generated from the meeting are properly allocated to a specific person, who is told precisely what is required, with a clear deadline. Appendix 2 (*Effective allocation of tasks*) describes an approach to allocating and receiving tasks in a working environment. As a bare minimum, all action points should be noted by the person responsible for the minutes.

The minute taker

In this capacity, it is important that you understand right from the start what the required output of the minutes is. It may be that it is purely a file note for those attending, where just broad highlights are noted. At the other end of the spectrum it may be a complete account of everything that was said and by whom. Either way, when taking the minutes mark contributions with people's initials in order to remind you who said what. Whatever the level of detail required, at the very least you need to document all of the action points allocated to those attending. In effect this means taking a note of what needs to be done, by whom and by when. In this respect you are also a buffer for the chairperson, helping to ensure that tasks have been properly allocated. While you are taking the minutes, it would normally be perfectly in order, if you notice that a task isn't properly allocated, for you to seek clarification on what is intended.

In a similar vein, especially if your role is to take extensive minutes, you need to be able to apply common sense as to what precisely needs to be recorded and what is not appropriate to include. Bear in mind that once typed up this document may be referred to again at a later date, or read by people who were not present at the meeting. Therefore, judgement and sensitivity is required. Occasionally you may find yourself in a position when you are not sure whether to minute something or not. Simply ask the chair if they want it recorded and proceed as requested.

If you know in advance that you are going to be taking minutes, it is a useful approach to get hold of an electronic version of the agenda and print it off with large spacing between each of the agenda items. You could also insert what, who and when headings under each item to keep you appropriately focused as tasks are allocated. Having done this, you now have a template for recording information during the meeting. If you are also expected to have input to the meeting in any way beyond minute taking, I would not recommend typing the minutes during the event. Firstly, you need to be alert as to what is going on, and secondly, irrespective of your anticipated input, it will be a distraction for those attending.

After the meeting has concluded, the minute taker will need to arrange for the minutes to be passed out to all of those attending, along with copying in other interested parties who, for whatever reason, either couldn't attend the meeting or were not required to attend. In effect, like many others in the room that day, the minute taker has an action point: to produce the minutes, in the prescribed format, by a certain time or date.

Hopefully your notes, as taken during the meeting, will be clear enough for you to produce what's required quite easily. I would strongly suggest that you do this as soon as possible after the meeting, while everything is still fresh in your mind. Also, bear in mind that when the draft minutes go out to the other attendees you want those people to be able to remember clearly what was discussed and agreed. Therefore, generally speaking, it is not good form to pass out the draft some weeks later. In doing so you are making the job harder for yourself, as well as for others.

A good approach is to block the meeting off in your diary as being longer than it actually is, and when you get back to your desk, immediately after the meeting, continue as if you are still on meeting duty and do the minutes there and then. If you have adopted the approach I suggested earlier all you need to do is open up the agenda document that you amended in order to write the minutes and start typing in the key messages and action points. Unless agreed otherwise, the minimum that needs to be documented within each of the items consists of the action points, which should include what action is required, the name (initials would be quicker) of whoever is responsible, the agreed date for completion, and the name of the person to whom it should be delivered.

Having sent the minutes out in draft format, what you are next seeking to do is establish that everyone who attended the meeting agrees with your version of what occurred. This of course includes all of the action points assigned to the relevant individuals. You are now in task allocation mode yourself. The email you send out could be along the lines of that described in Case study 6.1.

Having received the responses to your draft, you would then go on to prepare the final version of the minutes and send these out to the same group of people as before, but also, if appropriate, at this stage copying in the additional interested parties.

The final document then becomes a working document that can be used by the chairperson or other line managers to monitor that the various action points are being completed as required. Additionally, if there is another meeting involving the same group about the same subject, then these minutes can be referred to in order to remind everyone what was previously discussed and agreed.

All attendees

'*Why am I here?*' The number of times I have asked myself that question! Seriously though, if you are at a meeting and you are asking yourself this, then there is probably one simple answer. You are there because you allowed it to happen. You agreed to go, or you didn't think that you had the ability to decline the offer or send someone else more appropriate in your place. OK, granted, there are occasions when you

Case study 6.1 Issuing draft minutes

Setting: A meeting took place earlier today and the minute taker is sending out a draft version of the minutes to those who attended. At this stage other interested parties would not be copied in.

To: All those who attended the meeting, including the chairperson

Subject: Draft minutes – team meeting, 32nd Neverember 2051

Good afternoon,

Thanks for your input into the team meeting held earlier today.

As requested I have now had the opportunity to complete a draft version of the minutes from the meeting. These are attached to this email. This draft hopefully covers everything that needs to be documented, including the various action points allocated to those concerned. As such, please note that there is a strong likelihood that the attached draft document includes action points that you are responsible for delivering upon.

Can I ask that each of you review the draft by close of play tomorrow and get back to me by email with any required amendments? Alternatively, if you are happy with what has been documented thus far can you please email me to confirm that this is the case.

Note that the finalised minutes, taking on board any suggested and agreed amendments, will be issued first thing Friday morning.

Regards

Robert

don't really have any choice. Like it or lump it, you have to be there and just get on with it. However, when you really think about it, you may find that there are many other situations where you could be excused from attending, or could send a substitute.

If you weren't sure why you were needed in the first place, then you should have made enquiries. It could be that you have totally missed the point as to what is actually going on, in which case you need to be there after all, or maybe your boss hasn't really thought it through and inadvertently is making you spend your time unproductively. This being the case, you might be able to remove yourself from the proceedings. Better that than you being an idiot for agreeing to be there and then having to regret the lost time against a looming deadline that is growing arms and legs on your desk.

There is an expression that I find I constantly say to myself. *'No benefit, no point.'* I have said this for so long that I can't honestly remember where I first heard it. I am not smart enough to have come up with it myself, I'm sure. It focuses the mind on what's important. OK, sometimes, politically, you just have to show face in order to create a good impression – but in that scenario you have a purpose to your presence (to create a good impression). Ask yourself the question. What is the benefit to me in doing this? Challenge yourself and others (in a friendly and constructive manner) as to the best use of your time in order to achieve what the business needs from you. Don't just adopt this for meeting scenarios, adopt it across all aspects of your professional life. Don't just aimlessly go with the flow and then find that you are under even more pressure to get what's really urgent and important delivered.

Having established that there is a benefit to attending the meeting, and that you are indeed going to be there, what should you now be considering? Table 6.2 gives you some thoughts, including a recap on some of the points I have already discussed.

Once the meeting has concluded, if you have been given an action point to deliver then it is your responsibility to see that it is carried out within the deadline agreed. If after the meeting you establish that there is a problem in doing so then you must make everyone concerned aware of this and agree an amended action point.

If there are any action points that you are responsible for delivering, you would be misguided to wait for the draft or final version of the minutes to be produced before taking meaningful action to start working on what is required. You know what is required. Schedule it in now and get things moving in the right direction. Naturally, you should also carefully check the minutes to ensure that you have not missed something during the meeting, or that someone else's action point has not been mistakenly assigned to you.

Formal contributors

As someone who is contributing formally towards the meeting, you have some other points to consider over and above what is required of an attendee. The main thing is to properly understand the objective and what part you are being asked to play (i.e. what's expected from your contribution). Sit down and really think about what you are delivering, the manner in which you are going to get the message across, and why it is important. If you are struggling to understand what's required then perhaps you need to ask someone (such as the chairperson) their thoughts on the matter (ask, don't stall). Also, if you are going to be presenting, it would be silly, to say the least, for you not to know in advance everyone who was going to be there. If it isn't already obvious to you who is going, then ask. You can only imagine the things that could go wrong and the risks you could expose yourself to otherwise. For

Table 6.2 Meeting attendance

Point to consider	Additional thoughts
Why is my presence required?	Understand why you are being invited.
	What is your role?
Am I the best person to be attending?	Can I send someone else in my place?
	It could be that there is someone you know who can contribute more effectively than you.
	It could be treated as a development opportunity for someone junior to yourself, as well as freeing up your time to work on something else.
How am I getting there?	Think about alternatives.
	Train as opposed to car may allow you to get work done on the journey.
What time do I need to leave to ensure that I arrive on time?	Give yourself plenty of time to get there on time.
	Your late arrival at a meeting is rude at best, and unproductive for everyone else who is there waiting on you.
Read the agenda and previous minutes	Read the agenda before you attend the meeting.
	Also refresh your mind about what happened at the last meeting by reacquainting yourself with the minutes from that occasion.
	Did you complete all of your action points?
Introducing yourself	Keep it short and to the point. Name, department and reason for being there.
Notebook, pen and diary	Always. Every time. No excuses.
Mobile phone	Switched off and out of sight. You won't be using it.
Don't bog things down	Avoid going into too much detail about specifics.
	Take a higher-level approach to describing solutions, etc.
	Avoid talking for the sake of talking (e.g. repeating in a different way what someone else has already contributed).
Stay on topic	Don't side-track the topic being debated or go off on complete tangents.
	No side talk/whispering.
	Respect and listen to everyone's thoughts, and don't interrupt others.
Select an appropriate seat	Get in early and choose a place to sit that reflects how you intend to interact during the meeting.
	Notice where others choose to sit and whom they choose to sit next to.

example, suppose you have put together a presentation that says something negative about another part of the business, only to find, when it's too late, that someone from that area is attending that day as a guest.

It is important that you have your piece of the meeting properly planned out. If it's a presentation, then ensure that it tackles precisely what needs to be discussed and ensure that it can be delivered in the time allocated, allowing for discussion time during and/or after the presentation. To run over your allocated time is disrespectful to others who have also got material to present. If you do run over, someone else's session will have to be cut short or the meeting will last longer than originally intended. In the latter scenario, not only have you been disrespectful to the other presenters, but your actions have had an impact on everyone attending. Also, consider whether or not it might be helpful to have a copy of your presentation printed off for attendees. Do you need other supporting material in order to help the points being made come across more effectively? Think what people might need to refer to during your session. Having done this, then please ensure there are enough copies for everyone.

SPECIFIC MEETING TYPES

I will conclude this chapter by providing some additional guidance and thoughts about certain types of meetings that ecologists may regularly find themselves involved in. Please bear in mind that all of the generic rules I have discussed so far will still apply, but with the additional considerations as described. In Figures 6.2 to 6.5 I tackle each of these scenarios purely from the perspective of an attendee, as opposed to the perspective of the person who has called the meeting or is managing the meeting or acting in a role similar to a chairperson.

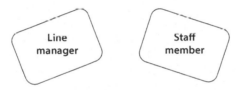

Line manager

Staff member

This is a meeting where you will be told how your performance has been over the period. Your manager will have looked at examples of your work outputs, made enquiries about your performance with others, and obtained the factual aspects of your performance as presented within management information systems.

You may very well have had current-year objectives (CYOs) that you were expected to achieve. If you have been smart you will have been monitoring your own progress against each of these CYOs, and every time you did something that would count as a positive towards completing a CYO you will have taken a note of it at the time.

You should not need your manager to tell you about the factual aspects of your performance. It is surprising how often employees go into these meetings, which are probably the most important internal meeting they are likely to have that year, having done no preparation whatsoever and having no examples or figures to hand.

Usually you would expect your line manager to have reviewed all of the documentation relating to any 121s, mid-term reviews, etc. as part of their preparation for the performance review. You should adopt a similar approach.

Figure 6.2 Appraisal/performance review with line manager – points to consider

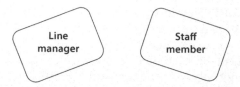

Understand the meeting structure that your manager may be using. If you understand better how these meetings are structured then you are likely to be far more effective. This is your 121, and you are allowed to introduce items onto the agenda.

Resist the temptation to butt in. Allow the manager to explain precisely what they mean by things they are discussing with you. Stay positive throughout and view the session as an opportunity to get a feel for how your manager is viewing your performance and in what respects you are doing well and what you need to work on.

If you are being given negative feedback, don't panic. Resist making excuses and do take on board the advice. If feedback hasn't been forthcoming, then ask for it. No feedback (bad or good) can equate to stalled developmental growth on your part. Refer to Appendix 1 (Feedback: get rich quick) for more guidance.

Figure 6.3 121 or feedback session with line manager – points to consider

Case/project ecologist   Case/project manager   Minute taker

Organise thoroughly and be on your most professional behaviour. This is the time for you and everyone else attending to impress a client and create the confidence they need to have in your ability to deliver.

Never criticise others in your organisation in front of a customer and don't disagree with your line manager or your project manager in front of a customer.

Don't make promises you can't keep.

Customer / supplier representative   Customer / supplier representative

Figure 6.4 Customer meeting – points to consider

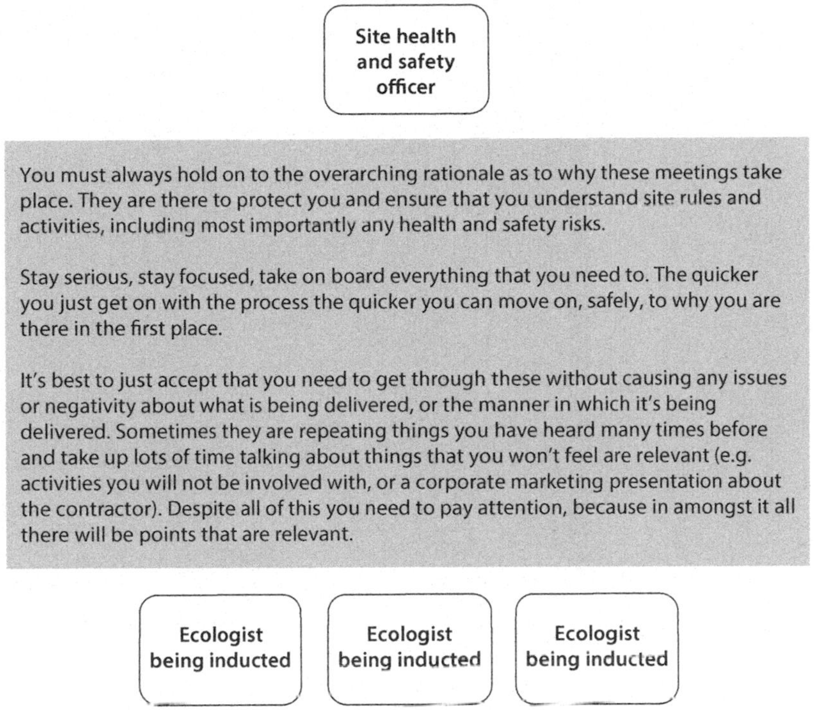

Site health
and safety
officer

You must always hold on to the overarching rationale as to why these meetings take place. They are there to protect you and ensure that you understand site rules and activities, including most importantly any health and safety risks.

Stay serious, stay focused, take on board everything that you need to. The quicker you just get on with the process the quicker you can move on, safely, to why you are there in the first place.

It's best to just accept that you need to get through these without causing any issues or negativity about what is being delivered, or the manner in which it's being delivered. Sometimes they are repeating things you have heard many times before and take up lots of time talking about things that you won't feel are relevant (e.g. activities you will not be involved with, or a corporate marketing presentation about the contractor). Despite all of this you need to pay attention, because in amongst it all there will be points that are relevant.

Ecologist
being inducted

Ecologist
being inducted

Ecologist
being inducted

Figure 6.5 Health and safety site induction meeting – points to consider

Chapter 7
PROJECT MANAGEMENT

To fail to plan, is to plan to fail.

Benjamin Franklin (1706–1790)
A Founding Father of the USA

Project management is a large and complex subject and could easily take up a whole book on its own, as demonstrated by the vast amount of literature available elsewhere. In this chapter I am going to approach the subject from the viewpoint of what would be deemed, in project management circles, relatively straightforward projects. As such, I need you to be aware that there are a number of areas relating to this topic that I won't cover (such as critical path analysis, risk management and project finance). I am going to restrict the discussion to those processes and tools that are relevant to an ecologist's day-to-day projects. If you are ever going to be involved as a project manager (from now on referred to within this chapter as a PM) developing and overseeing a longer-term complex project involving many people and skills beyond your area of technical expertise, then please do appreciate that there is more to this subject than what I have included here.

As an ecologist, the projects you look after can be relatively complex and, accordingly, knowledge about project management techniques can help you considerably. Within our sector ecologists are regularly involved in delivering one part of a much larger project, and in many cases, although you might perhaps view the piece you are involved in as a project in its own right, the work you are responsible for is purely one element (the ecological components) of something much bigger.

Now you may be thinking that a lot of the work you are involved with isn't really at such a large or complex scale that it would warrant being called a project in the first place. For example, is a single badger survey of a piece of woodland a project? Well, compared to other pieces of longer-term and multifaceted work (such as a suite of ecological surveys being carried out over 12 months as part of a new road scheme) it clearly isn't on the same scale. But, whatever the scale of the work you are involved with, it is certain that many of the project management techniques that I am going to discuss could be of great assistance. So rather than ask, *'Am I involved in a project?'*, ask yourself this: *'Can any of what follows be applied to any of the tasks that I am involved with, irrespective of size or complexity?'* I think you will find that, very often, it can.

What is a project?

A project can be large or small, and can take place over a matter of days or a considerably longer period of time. It may be a fairly straightforward process or something quite complex. It could be purely 'in-house' (for example, implementing a new IT system or the development of a new product), or alternatively, the people requiring the project to be delivered could be external to your business (in other words, a customer). For most ecologists, by far the bulk of their project work is going to relate to the latter (a customer requiring ecological studies or expertise at some level). Therefore projects don't need to be large. In fact anything involving more than one task, or more than one person in its delivery, could be deemed a project. So now you are thinking, *'Well that's pretty much everything then, isn't it?'* The answer: *'Yes, it is!'*

Broadly speaking I would say that a project can be defined as follows: A piece of work that is going to take longer than a day to complete and which typically relies on at least two people delivering key components in order to satisfy the overall objective within the required time frame, to the required standard and within budget.

Now you may be thinking, *'Brilliant, if it's just me responsible for doing a straightforward Phase 1 report for a small site, that's not a project and I can skip to Chapter 8.'* I, on the other hand, would argue that it still fits into my definition. What about the logistics involved in accessing the site? Do you need to involve someone else? What about the desk study? Do you need to approach the local records centre? What about the quality of your report? Does someone need to proofread it? There you have it. It's not just you that is involved after all. There is a reliance on communicating with others in order to achieve the desired result, and as soon as you have more than one person involved, communication needs to be effective. There are a number of component tasks involved: the research ahead of the site visit; the field survey; producing the report. You put all of these things together and project management skills are certainly going to help. Chapter 8 will have to wait a little longer.

Project management skills

In order to give yourself the best chance of achieving the result that your customer is expecting, you would be well advised to adopt a project management type of approach to the work you are tasked with delivering, irrespective of size or complexity. At times it may be that you don't need to use all of the project management techniques in your toolbox, but at least it is good to know that you have them there, ready to be taken out and deployed if required.

In order that projects can be effectively delivered there are a number of steps you can take to help ensure that it all happens as envisaged. Figure 7.1 lays out the key components involved in successfully managing a project.

First of all you need to fully understand the objective, and having done so appoint a suitable team to develop and implement a plan of action. During the process there will be a need to collaborate beyond the project team, including of course with the

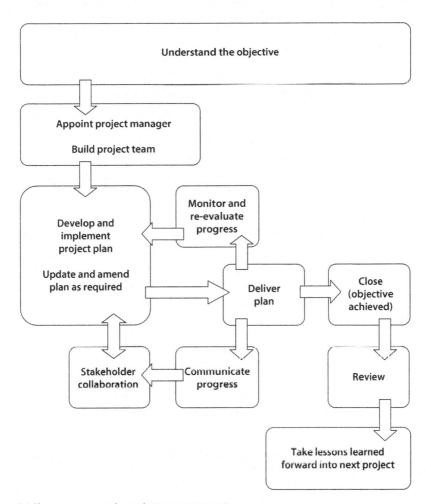

Figure 7.1 Key components in project management

customer. Progress will need to be monitored, and if required amendments to the plan will need to be made as matters progress. Finally, you will reach completion and the project will be closed off. But this shouldn't be the end. One last vital step should be to review what has taken place. What went well? What could have gone better? What did we learn? It is important to take these key learnings forward so that you, the company you work for, future customers and their projects can all benefit.

From conception all the way through to closure there are so many things that could go wrong, even in the smallest of projects (imagine that you have two days to turn around an otter survey and a storm is in full throttle, going nowhere fast). Understanding and effectively managing the project is essential. It needs to be delivered on time, to the required standard and within budget (Figure 7.2). At every stage, as you progress towards achieving the objective, you will be relying on other

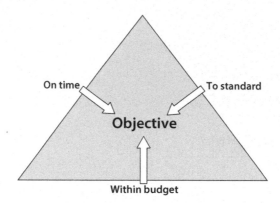

Figure 7.2 Key high-level elements contributing to successful project delivery

people. Are they capable, are they on board, are they interested or committed to the same level as yourself? The potential for others not doing what is required of them is always there. It needs to be managed, and if things don't happen as they should, you have the ever-present enemy, time, working against you. Tick, tock: everything is taking longer and being pushed back, except for the deadline – that hasn't moved. You need to be on top of all of this and so much more. It needs to be managed. Yes, I am repeating myself. After all, repetition is reinforcement.

In my business we have borrowed an expression that we apply to managing projects: *'You need to be all over it like a rash!'* I think that describes it perfectly. You need to know about, be close to and be in control of everything that is going on. I am now going to cover, in more detail, the key components as described in Figure 7.1.

UNDERSTANDING THE OBJECTIVE

Before you can begin to start managing any project you need to fully understand the objective. In order to do this there are many things that should be explored. Let's start with, who is your customer? You need a level of understanding about them, to a point where you can see why this project is important and how it fits into their wider goals. Understanding who your customer is and having a developing relationship with the key players within their business could prove very useful during the lifespan of the project, and possibly beyond this, as other projects come on-line. Turning this around, how dangerous would it be if you didn't understand your customer and if you didn't have any relationships developing with the key people involved in achieving their goal?

Having established good customer knowledge, next you really do need to understand, as thoroughly as possible, what precisely it is that they require (i.e. their objective). If you don't know precisely what the objective is, how do you know where you are supposed to be going and when you have finally arrived? It is not good enough to be told simply, *'We want to build a hotel.'* You need to know what stage in the process the customer is currently at. When do they anticipate needing your component part of their wider project plan delivered? What precisely are you required to do, and why is it important? How does your component part impact upon the rest of their plans? What does the result they are looking for, from you, look like when it's completed? Is there a budget, and if so what is it? It may be that your budget has already been agreed and capped at the figure quoted while acquiring the business. There is certainly a benefit to having all of these areas understood at the very start, and you can never have too much information. Having had these discussions, it is essential for the agreed brief of the project to be formally documented in order to ensure that there has been no misunderstanding as to what is required. In addition, a documented brief also acts as a point of referral throughout the life of the project, in order to ensure that the project team stick firmly to what was agreed at the outset.

For larger or more complex projects you will often find (in fact it would be worrying if you didn't) that all aspects relating to the project are fully documented. As an ecologist involved with a component part you may not be aware of or see this detail, but if it exists there may be a case for at least some aspects of it being made available to you. It depends on how involved your deliverables are to the overall project. If it's just a case of *'We need a bat survey of that structure within the next eight weeks'* you may very well have enough information at your level, provided of course you stick firmly to the brief. At the very least, however, I would want to know why the eight-week turnaround was important and what would be the consequences to the project overall if this deadline wasn't met? If, on the other hand, you are going to be involved with a broad suite of work in connection with a site over a longer period of time, then I would strongly suggest that you should request as much additional information as is available.

PEOPLE MAKE IT HAPPEN

Nothing happens unless people make it happen, and as PM you essentially have three groups of people to consider. First of all there is the project management team that will be in place in order to deliver the required objective. Then there are the two groups that collectively form the external stakeholders. Usually one particular stakeholder stands out from the crowd for obvious reasons, and that is the customer. Finally there are all the other stakeholders, external to the project management team and the customer, who could potentially have a positive or negative impact upon the project along the way. I will now tackle each of these three groups separately in more detail.

The project management team

Quite often people talk about a project as if it's a thing, a large mechanism without any personality or emotions that is driving us all bonkers. *'No one is to blame. It's just that this project is a nightmare.'* ... *'The quicker I can get away from this project and on to the next one the better.'* The point that sometimes gets lost in amongst all of this noise is that the project delivery process is all down to people. These people, at whatever stage in the process they are involved, have agreed that the objective is deliverable, and have then set in motion the process to ensure that it is going to happen.

With all that in mind, there is a strong need to ensure that the right people are in place to make it all happen, and that these people have the appropriate time and resources. So, who are the right people? Well, that of course varies depending upon what aspect of the plan you are working on, and what skills and resources are required along the way. It all starts with managers at the very beginning of the process appointing the correct people to deliver what's required. The first and most important appointment is that of the PM. The managers may then leave the PM to build the rest of the team (from here on referred to as the project management team), or alternatively they may still be involved in the appointment of others working beneath the PM.

There needs to be strong leadership, drive, commitment, organisational skills, people skills, negotiation skills, and so on ... Now sometimes that just isn't there at the very start and the person appointed as PM to drive the whole thing forward is put in place because, for example, they have strong technical skills. If, therefore, the person doesn't have the experience or skills in running a project, it quite possibly is going to go belly-up right from the start. And who is to blame? It's all too easy to blame the PM. I would strongly suggest, however, that if someone is appointed to that position and they don't have the skills or haven't been given the support and training to carry out the role effectively, then the fault lies with those who made the appointment in the first place. Therefore, it is down to the people at the top making the right judgements and decisions from the get-go. The key thing is to have the right person in charge and then give them the right level of support to put a good project management team in place, all in order to get the job done. This project management team is not, in the context of what I am describing within this chapter, purely a group of 'managers'. They are the team of people who have been put in place to deliver to the requirements of the project plan, as opposed to solely managing people, albeit within the team there will be people performing conventional 'managerial' roles (someone needs to be in charge of what is happening).

In appointing the PM many things will need to be considered. What is their degree of experience handling projects similar to what is involved on this occasion? Are they experienced at project managing? At what level does their technical knowledge 'kick in' for the tasks they will be managing? As well as managing the project, can they manage people? Are they experienced and knowledgeable about the financial aspects and managing a budget? In regards to all of these questions, do they need technical, human resources or finance support, for example?

It would be not be unusual, during the early discussions with a potential client, for the earmarked PM to be identified and included in the meetings when the brief is discussed. If this hasn't been the case up until now, then once the project has been given the green light to proceed, a PM needs to be put in place immediately. The key to all of this is that the PM is going to need to be suitable in relation to the customer and the objective sought. The customer will be seeking reassurance, at some level, that the PM, who will ultimately be accountable for the delivery of the project, can deliver.

Once in place, the PM would then begin to build a project management team around them who are able to take on responsibility for many of the tasks, which in combination will take them through to the delivery of the overall objective.

Some of the project management team members may be identified and brought on board very quickly, while others may join the team later on, or at certain points as things develop. These team members may be focused full-time on this one project, or they may also be carrying out other unrelated work elsewhere in the business. However it comes together, the project management team needs to be created with the objective of the project in mind. To give you an example – if part of the project involves the production of an ornithology chapter within an ES (environmental statement), you will need an ornithologist experienced in this type of work within the team, as well as possibly additional field ornithologists to help out with the surveys. If you know this from the start (which you would do if the brief has been fully discussed and developed) you should have these people identified, lined up and ready for when you need them. You shouldn't be scurrying around at the last minute looking for people that you have known right from the beginning will be needed.

As the team begins to take shape it is a good idea, as far as practicable, to get everyone involved in the development of the parts of the plan that they themselves will be responsible for. When people have been included in this way you often find that the 'buy in' to the process and the commitment to deliver is at a much greater level. Accordingly, as PM, let the person you have appointed to lead the bat surveys decide specifically when the surveys will be done, the application of the methodology, and who they will have in their survey team. Provided, of course, that they keep you in the loop and can demonstrate that what they are doing works and that they can deliver what is required on time, at the right standard and within budget.

Once the PM and the rest of the project management team are in place, it is also essential that the customer is made aware of who they are dealing with, and what the various roles and responsibilities are within the project management team.

The customer

Bearing in mind the overall objective of the project and the importance of it being delivered, the customer is, with very rare exceptions, the most important external stakeholder that the project management team will need to consider. It is the customer who is paying for the delivery of the objective and it is the customer who probably has most to gain or lose on the project.

Understanding the customer, relative to the project, is essential. Your customer may, for example, be another firm of ecologists, or a big engineering company, or an architect. In such circumstances you may not actually have any direct involvement with the ultimate customer (whoever it is that your customer is working for). Nonetheless, in such situations you should still try to establish where you sit within the bigger picture. Although you may be a number of steps removed from the ultimate customer, you should not lose sight of the fact that your involvement is part of a broader project within which you are required to deliver the specific tasks assigned to you.

Whoever it is you are reporting to, at whatever level, you need to know who specifically you are dealing with. What is their role? Who deputises in their absence? Who do they report to? Are there other stakeholders identified at their end that also need to be consulted or kept informed? Who are all of these people they keep copying into the emails that they are sending you? Find out, and do it quickly, before it becomes awkward to ask. All of this detail will make life so much easier and more effective as matters progress.

Other external stakeholders

Beyond the project management team and the customer, there sit other people (other external stakeholders) who will have some sort of vested interest, either directly or indirectly, in the project. It would be risky to ignore their existence. In amongst this group there is the potential for issues to arise, or for scenarios to be dealt with less efficiently than they need to be. Poor management of these stakeholders could very well impact upon your ability to deliver what's required effectively. With a bit of research, and adopting a targeted approach to each of them, most of these stakeholders (if not all) can be managed by the project management team.

Very early on it is extremely important that the project management team establishes who these other external stakeholders are. They will be present in all shapes and guises. Most will be interested because they support what is going on, but there is always the potential for others to be against the aims of the project, either openly or working quietly behind the scenes in order to disrupt or derail the process.

If you are oblivious to these stakeholders you are, at best, risking upsetting some people along the way because you haven't paid them the attention they are due. At worst, you may be caught unawares when someone below your radar causes real damage to your progress. You need to identify, as far as possible, all of the stakeholders connected to the project you are working on, and having done so you must adopt suitable methods of communicating with and/or involving them with what is being delivered. Some of these stakeholders will need to be heavily involved at times, as you and the project will benefit from their contributions. There will be others who may not contribute much, if anything, but who nonetheless need to be kept informed as to progress. Establishing the attitudes of stakeholders can also help you identify people you may wish to involve more, or less. Having identified all of the stakeholders, you should now go on to document them according to their perceived attitude (positive, neutral or negative) towards the success of the project, for example as demonstrated in Table 7.1.

Table 7.1 Initial stakeholder analysis: example

Name and position of stakeholder	What is their interest? or What do they require from the project?	What does the project require from them in order to achieve the objective?	Perceived attitude (positive, neutral or negative)
John Smith (JS) Managing Director A B & C Ltd	Tenant within the existing factory unit They will be expanding into the ground floor of the new premises Important client to landowner	Access to the survey site through their gated premises	*Positive*
Martin Jones (MJ) Chairperson Local Badger Group	Badger sett in adjacent land to new site Concerned that badgers will be impacted by work	Historical data regarding the sett (location, usage, etc.) Site meeting to discuss location of sett relative to work	**Negative**
Penny Sterling (PS) Head of Accounts Department	Responsible for processing income and expenditure items	Efficient processing of invoices Ensuring suppliers are paid Credit control Financial reports	Neutral

Having done this, you can now go on to organise the information in such a manner that you can best decide how and towards whom your efforts are most effectively focused. There is a well-established, industry-wide 'stakeholder management model' that you can use to assist you in achieving this (Figure 7.3). There are many styles to this matrix, so bear in mind that what is shown in Figure 7.3 is just one version of how the information can be organised and displayed.

Allowing for the stakeholders' perceived attitudes, you would plot each one's position of power/authority over the success of the project, against their level of interest. In doing so, if you also allocate a different colour code for each stakeholder according to their attitude ('positive', 'neutral' or 'negative'), this enables the data gathered in Table 7.1 to be retained alongside the information plotted in the matrix, as demonstrated in Figure 7.3.

Using the examples from Table 7.1, John Smith (JS) would be regarded as having a very high degree of interest in the project, but not so much power/authority. As such, he would be plotted just within the higher box on the right-hand side ('involve'). Considering Martin Jones (MJ), we would probably determine that he has a high degree of interest, but no real power in the grand scheme of things. Accordingly, he would be positioned within the lower box on the right-hand side ('acknowledge').

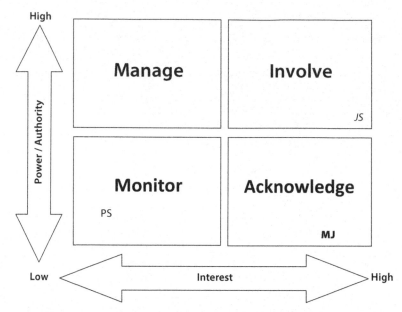

Figure 7.3 Stakeholder management model – power/authority against interest. Note that in place of colour-coding, this black-and-white version of the matrix uses *italic*, **bold** and normal text for the stakeholder initials to indicate *positive*, **neutral** and negative attitudes.

Finally, Penny Sterling (PS) would not have much power over the project and probably not a huge amount of interest. She would be placed in the lower box on the left hand side ('monitor').

Taking on board all of the data you have now gathered, the key is to implement a plan of action, and appropriate techniques, targeted towards any stakeholders who can either be engaged in order to support the success of the objective, or who can be managed in order that the impact of their negativity is reduced or eliminated. For the latter, if they are initially negative about what is happening, is there something that can be done to draw them towards a more neutral position? Or, even better, can you turn them around completely to the point that they become an advocate? This may seem unrealistic, but negativity may stem from not being fully aware of everything, and once the full story has been communicated a complete change of attitude can occasionally occur.

A substantial risk to the project is present when people with high degrees of power or high degrees of power/interest combined are not fully on board. If you have any doubts as to where to concentrate most of your efforts in handling stakeholders, then most definitely it would be with those. The model allows you to easily see who they are and points you in a direction as to how you should deal with them. Table 7.2 provides some guidance in this respect, showing how the labels in the four boxes of the matrix (Figure 7.3) can be used as the basis for developing approaches to manage stakeholders. While doing all of this, please also bear in mind

Table 7.2 Stakeholder management

Position within model	Management style	Additional notes
High power High interest	Involve	This group potentially has lots to offer, or alternatively could cause huge disruption at a high level if not catered for. It would be sensible to communicate with and involve them constantly throughout the life cycle of the project.
		If they are in a 'negative' or 'neutral' attitude category you should really be making concerted efforts to turn them to 'positive'. The time taken should be worth the effort.
High power Low interest	Manage	Lots of authority but less interested specifically in the project, albeit they may need to be catered for and included at certain times. Invite them to attend key events and meetings. They may or may not show up, but being invited is the key to keeping them happy and supportive.
		If they are 'negative' towards the project, their high level of power is a potential problem that needs to be addressed.
		Due to their lower level of interest, they may be more 'neutral' in attitude. That may not be a huge problem provided they are not going to do anything obstructive.
Low power High interest	Acknowledge	They can't really do much harm in terms of their power, but they are very interested in what you are doing and may be able to influence others, positively or negatively.
		Acknowledge them and the areas they are influential within. Make them feel as if they are part of what's happening, but without involving them too deeply.
		If they are of 'negative' attitude, then being aware of that and trying to move them towards a more 'neutral' position would be beneficial.
Low power Low interest	Monitor	This group, at the outset, do not appear to have any position of authority that will impact upon the project and do not have a high degree of interest or influence upon others.
		You can proceed without having to worry too much, but monitor them in case things change. Individuals here may transfer into another box (e.g. through a change in position or change in personal responsibility, or if new information comes to light).
		If there is a 'negative' attitude within this group, do not be overly concerned and don't let it distract you from more pressing actions.

that during the life cycle of the project there is the potential for people to move, either in respect of their power/interest rating, and/or in respect of their perceived attitude. In addition, new stakeholders may emerge and conversely, some of those initially present may depart.

PROJECT PLANNING

Having established and understood the project objective and considered the people involved (in whatever capacity and at whatever level), you should now begin to develop a detailed plan in order to ensure that the project is delivered on time, to the desired specification and within budget. I cannot stress how important it is to spend time creating a robust project delivery plan. It is the first next thing you should do, and in the words of General George S Patton (US Army General, 1885–1945), *'A good plan today is better than a perfect plan tomorrow.'* Once the plan is in place you can always tighten it up, sort out more of the detail, or improve upon it as the project evolves.

A good tip is to take the overall objective and break it down into its main components, listing these components chronologically in the order that you envisage that things will need to be done. At this stage you do not need to be too detailed, but make a note of anything that comes to you so you don't forget in the future that which you have already thought about. Figure 7.4 demonstrates this part of the process, and during the remainder of this section I am going to take this fairly straightforward ecological example through a small project planning process in order to help demonstrate some of the techniques.

Figure 7.4 looks at a project in its early days of planning, when the PM (a senior ecologist, in this example) has been given a case in which they are accountable for

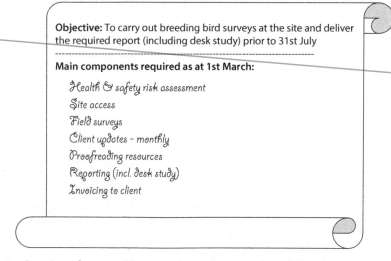

Figure 7.4 Overview of project objective with initial component tasks listed

ensuring that the work is delivered to the client no later than 31st July. You can see that even with this relatively straightforward project there are a number of components. As soon as you have as much as this happening there is a risk that somewhere, something gets missed or not done as quickly as it should have been. Therefore, there is the potential for the successful delivery of the project to be put at risk. If the project isn't delivered properly then it is less likely that you will have a happy customer. The perception of your line manager as to how effective you are at your job is also tarnished. Your colleagues may, as a result of your inability to plan properly, be put under unnecessary pressure, with all of the consequences that entails. I'll stop there. Hopefully you see why this is important. If you fail to plan, you most definitely *do* plan to fail. If you fail due to poor planning (or no planning whatsoever) it is most definitely no one's fault but your own.

Now that we have a list of the main components that need to be dealt with in order to arrive at the objective (Figure 7.4), the next thing you should do is allocate the specific tasks that need to be undertaken in order to support the delivery of each main component.

During this phase of your planning process you should be more detailed and identify as many of the time-critical aspects of each of the components and their associated tasks as you can. When describing specific drilled-down tasks you should ensure that the things that need to be done now, or shortly, are fully thought out and described. For tasks that are anticipated to occur later you don't need to be as detailed, as this may be wasted effort owing to unanticipated additional factors having to be considered nearer the time. The main thing is to think through each task and then to establish what needs to have occurred before you will be in a position to start that task. It is rather like playing a game of consequences, but working backwards.

When documenting these specific tasks you should also now assign deadlines to each. Referring back to the technique described in Chapter 5 (Table 5.3), start at the end of each process and work your deadlines backwards. As you set the deadlines you may encounter scenarios where more than one deadline needs to be allocated within a component or a task. This is easy to deal with by simply expanding the list. Table 7.3 shows what this would look like for the example being worked through.

It is very important to apply deadlines that are not the absolute latest that you can afford things to be completed by. Ideally you need everything to be planned so that it can be completed ahead of schedule. This approach is beneficial as it allows for any slippage that may occur due to unexpected occurrences. Also, it means that it is less likely that anyone else you are relying on ends up being given things to do at the last minute because of your bad planning. The proverb *'Poor planning on your part does not constitute an emergency on mine'* very much comes to mind here.

Your original component list (Figure 7.4) now has individual tasks added to it (Table 7.3), and it may also now be reordered from what you originally thought, as more detail regarding the specific tasks and deadlines has brought to light aspects that perhaps you hadn't initially considered. During this second part of the process (the specific task allocation) you have been able to test and, if required, reprioritise the chronological order that you originally envisaged.

Table 7.3 Identify the tasks within each component and apply deadlines accordingly

Component	Specific task	Deadline
End point 6. Invoicing	Invoicing to client: two invoices to be issued as per terms of agreement.	31 July 14 June
5. Reporting	Corrections made. Final report signed off and issued to client.	28 July
	Draft report completed and proofread by line manager.	7 July
	Reporting template – initial set-up and complete general areas, desk-study results and methods.	30 April
	Deadline for responses.	14 April
	Desk study: send data request to local records centre and raptor group.	31 March
4. Client updates	Client updates: monthly progress updates by email.	31 March onwards
3. Complete field surveys	Breeding bird surveys – field work to commence by 28 March. 1 surveyor/4 visits (3 dawns plus 1 dusk).	28 June 28 May 28 April 28 March
2. Site access arrangements	Advise client/adjacent landowner of proposed survey dates and then remind them two days in advance of each survey, with relevant details regarding people present and vehicle.	21 March
	Speak to client regarding padlock code for locked gate, for vehicle access.	7 March
	Name and contact details for adjacent landowner.	7 March
1. Health & safety risk assessment	Review and amend after first site visit. Line manager to sign off. Submit copy of final version to client.	14 March
Start point	Prepare draft before first site visit. Line manager to sign off.	7 March

You have now identified what needs to happen, both at a higher level and at a more specific task level. You have also established when each of these specific tasks needs to be completed. The one thing that is missing, up until this point, is who precisely is going to do all of these things. I am now going to introduce another well-established, industry-wide model that can help you with this: the RACI model.

The RACI model

As a PM you need to be able to identify who is going to perform each specific task. In addition to those responsible for the tasks, who else needs to be consulted or communicated with as to progress being made and the eventual completion of the task? How do you interact with and manage all of these people in terms of what the project requires from them, and what they might require from the project management team? All of this is required in order to arrive at a successful result.

The RACI model can be used to clarify how the people fit in to the project plan at every level of the project right down to individual tasks. RACI stands for 'responsible', 'accountable', 'consulted' and 'informed'. The model takes each of the tasks that make up the project and asks you to establish who is responsible for the task being completed, who is accountable for ensuring it happens, who needs to be consulted before or during the process, and, finally, who needs to be kept informed or updated along the way.

Adopting the RACI model allows you to ensure that the different people involved, directly or indirectly, and also those acknowledged as stakeholders, are always considered. An important thing to remember when using the model is that everyone is considered against each specific task, and therefore people may be categorised differently in relation to other tasks: for example, someone may be 'responsible' for the completion of one task, but placed in the 'consulted' category for another. It is also worth bearing in mind that as well as individuals you can allocate role descriptions to departments and the like when applying RACI. The only time you cannot do this is when allocating 'accountability'. Accountability must always be allocated to an individual. Table 7.4 provides some more detail of the RACI model.

When you are involved in a project, apply the RACI model and establish whether or not you are responsible or accountable for what is happening at any particular point in time. Having done this, establish who else is responsible for any of the deliverables and who may be accountable for your responsibilities. Finally, who are the people that you need to consult with, and who are the people that need to be kept informed regarding progress?

If you are the PM, your line manager will have assigned responsibility for the delivery of the work to you, and the line manager will in turn be accountable for ensuring it all happens. When it comes to your responsibilities, at PM level, you seldom will be able to do it all on your own. You will need assistance from quite a few others. Some of these people will already work within your business (e.g. the GIS technician), and indeed in your team (e.g. the ornithologist), while others will be external (e.g. the local records centre) and, as such, not so easily managed, influenced or controlled. Within your role as PM you may pass some of your responsibilities on to others at certain points of the process. For example, you may not be involved with the field visits for the ornithology. On those occasions the surveyor who is sent out to do the work is responsible for those specific tasks on those days, and you are accountable for the delivery.

In regards to all of these other people who are going to be involved in your project management team, you need to get them all on board as quickly as you can in order to begin to make things happen. Once each of them understands what their responsibilities are, they can then start planning and prioritising their time in order to deliver what is required within the deadlines you have prescribed.

Table 7.4 Allocating task involvement using the RACI model

RACI descriptors	Guidance notes
Responsible	Those who are physically responsible for delivering what is required.
	An everyday ecology example could be an ecologist charged by their line manager to carry out a specific bird survey on a specific day. The ecologist is responsible for the delivery of the required result.
	At a task level it is possible for more than one person to be given the responsibility, or for the responsibility to be allocated to a department or organisation.
Accountable	Those who are accountable are in a position of authority over those responsible, and as such are held fully accountable for ensuring that everything that is expected of those responsible actually happens.
	This group do not carry out the work specifically, but must ensure that it is delivered.
	In our example the line manager allocating the bird survey to the ecologist would be deemed accountable.
	Note that, on occasions, it is possible for the same person to be both accountable and responsible: for example, if the line manager went out and did the bird survey themselves.
	Accountability must always be allocated to a single individual.
Consulted	This group may consist of people sitting firmly within the project management team, or those who are periodically attached to the team, or external stakeholders.
	Wherever they are positioned, it is agreed that their input at certain points (e.g. during specific tasks) is required. Usually the advice would be sought ahead of key decisions.
	In our example the land owner may need to be consulted as to what date is suitable for those carrying out the bird surveys to be on their land.
Informed	This group may consist of people sitting firmly within the project management team, or those who are periodically attached to the team, or external stakeholders.
	Wherever they are positioned, their input at certain points (e.g. during specific tasks) is not required, but they need to be kept informed about progress.
	These people need to be advised about what is about to happen or what has already happened. It is important that the lines of communication are such that they hear these updates via the proper channels and not from other sources.
	In our example the developer may be required to be informed as to how the bird survey programme is progressing and advised of any important discoveries made that could impact upon their plans.

Usually, the most sensible approach to getting things done, for anything and everything that relies on others, is to get these tasks off your desk and onto theirs immediately (Figure 7.5). As long as someone else's task is still on your desk it is going nowhere, slowly. Also, the quicker you can get all these other things moving, the clearer your mind will be in order to focus on the tasks that you have assigned to yourself. For those tasks that you are totally in control of (those that do not in any way involve anyone else), set aside appropriate time slots in your own schedule to have them completed ahead of their deadlines.

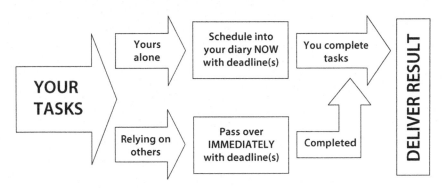

Figure 7.5 Task management flow chart

When allocating tasks to others, always adopt the task allocation approach as outlined in Appendix 2 (*Effective allocation of tasks*). In addition, when setting objectives and targets for others it is recommended that the industry-wide SMART model (Drucker 1954) is adopted (Table 7.5). SMART is a mnemonic that is used to remind us that objectives or targets being set should be 'specific', 'measured', 'agreed', 'realistic' and 'timed'.

Table 7.5 Setting SMART targets and objectives

S	M	A	R	T
Specific	Measurable	Agreed	Realistic"	Timed
What is specifically required? It needs to be clear. It cannot be vague.	What is the measure that is used that will show the task has been completed?	Has the person responsible for the task been told what it is, and have they accepted responsibility?	Is what is being asked realistic? Is it achievable in the first place?	Within what timescale, or by what date, does the task need to be completed?

*Note: It is perfectly acceptable within this approach to ask for something to be done that can only be delivered in a manner that stretches the individual or resources. Something would only be deemed unrealistic if it was impossible to complete even allowing for other work being reallocated or additional resources being made available, for example.

The Gantt chart

There are many project management software packages and tools available for those organising even the largest and most complex of projects. I don't plan to describe these in any detail (most of them I have never used myself, in any case), but I want to make you aware that there is software out there that can assist you greatly if your typical project is complex. What I am going to share with you, however, is an approach that often gets integrated within these software packages, called the Gantt chart. It is a fairly straightforward concept to grasp once you get into the swing of it, and it can be adapted in so many ways to assist with the overall planning and management of the typical projects that ecologists get involved with.

The Gantt chart has been around for over a hundred years and is probably the most powerful and frequently used approach there is for planning projects, whatever their size and complexity. It takes its name from a US engineer called Henry Gantt (1861–1919), who devised the technique in the early twentieth century. If you have never heard of a Gantt chart before then what follows could very well be a life-changing moment for you. You may have seen one during client meetings, or on the wall in a contractor's office. You may have even received one by email from a developer, as they share with you the plan as to what needs to get done, by whom and by when. Figure 7.6 shows the project example that we have been working through displayed in a Gantt chart, incorporating all of the information we have obtained. The Gantt chart shown in Figure 7.6 is only one example of how these charts may appear. There are numerous approaches that can be used, and lots of additional information can be incorporated. Usually they show tasks listed down the left-hand column and dates shown across the top row – though note that the version in Figure 7.6 has been turned 90 degrees anticlockwise so that it fits more neatly on the printed page. When completing task descriptions within a Gantt chart it is usually best to use a verb followed by a noun, as this denotes action required (e.g. 'create report template' or 'issue invoice').

For straightforward projects, you can usually build a Gantt chart quite successfully using software such as Microsoft Excel. But first, a health warning. You will be familiar, I am sure, with the expression *'rubbish in, rubbish out'*. We come across this quite often when dealing with data, statistics and software programs, and the same very much applies to developing Gantt charts. It's down to you, as the PM, to ensure that the information you put in is correct in the first place. Are the deadlines achievable? Are the tasks allocated to the right people? Do they have the resources in place to get the job done? It is therefore very important to think carefully and fully through everything that needs to happen in order to deliver the project successfully. You may need to gather people together and have a bit of a brainstorming session (see Appendix 3, *Effective brainstorming*). Many minds are better than one.

Within a Gantt chart you can have tasks that can run parallel to each other, tasks that overlap, and tasks that are sequential. When tasks are sequential, it is often because one task cannot start until another has concluded. It is important that you recognise this inter-dependability, as missing a deadline in one area could have a knock-on effect on other aspects of the plan. In other circumstances, missing a deadline may put certain aspects of a plan behind schedule, but allow other areas to progress as originally envisaged.

Deadline dates (When?)								
July	31							Invoice 2
	28						To client	
	21							
	14							
	7						To proof	
June	30					X		
	28			Visit 4				
	21		X					
	14							Invoice 1
	7							
May	30					X		
	28			Visit 3				
	21		X					
	14							
	7							
April	30					X	Template	
	28			Visit 2				
	21		X					
	14						X	
	7							
March	31					X		
	28			Visit 1				
	21							
	14	Final					X	
	7	Draft						

Ref	Task description (What?)	Person (Who?) (RACI)					
1	Prepare and sign off health & safety risk assessment	SE(A) EC(R) LM(C) CLT(I)					
2	Advise client of survey dates Sort out access	EC(A) AE(R) CLT(I)					
3	Complete field surveys (x 4 visits)	SE(A) EC(R) LM(c) CLT(I)					
4	Client updates	SE(A/R) EC/LM(I) CLT(I)					
5	Complete reporting and issue to client	SE(A) EC(R) LRC(C) LM/CLT(I)					
6	Issue invoices	SE(A) AE(R) FNC(C) CLT/LM(I)					

Figure 7.6 Gantt chart showing tasks requiring to be actioned in order to meet objective

The shaded boxes in our example are there to demonstrate when a task is anticipated to commence and when it is targeted to be completed. As progress is made, boxes are shaded in darker to demonstrate where tasks are progressing or have been achieved. I have shown some examples of this in Figure 7.6.

Gantt charts can often become quite overwhelming because of the amount of information they are required to convey. In these circumstances it is not unusual to have a higher-level chart showing the main components and, sitting beneath this, various subordinate charts for individual components expanded out into their individual tasks, and so on. For our worked example I have kept things fairly straightforward, but bear in mind that even the simplest of projects can have a large number of individual tasks that need to be completed in order to achieve the objective.

You will notice that some of the tasks in Figure 7.6 have an 'X' inserted against them. Here, another type of descriptor has been introduced: the milestone. Milestones are points (i.e. dates) within the plan when something is supposed to have occurred by. These would be regarded as significant points in the overall delivery. In contrast to tasks, milestones are often described by a noun followed by a verb (e.g. 'survey completed').

As it refers to something due to happen by a specific date, a milestone would normally be represented by a single point on the chart (in our example, an 'X'). You can adopt a similar approach to show when review meetings will take place and progress reports are due to be delivered. Within our example we have again purely shown these as an 'X' against the appropriate task. One final point to keep in mind is that quite often milestones and review points may sit separately within their own headed lists, beneath the tasks.

MONITORING PROGRESS

During the project it is important to monitor progress. How else would you know if you were on schedule, if you did not monitor along the way where you were in relation to the plan? You should be regularly referring to the plan, and if necessary making adjustments as you go. If you don't monitor progress you may very well struggle to see that the project is not on plan and miss the opportunity to take corrective action before things get out of hand.

Your team must feel confident that they can report potential delays and blockages as soon as these become apparent and without serious repercussions. Such scenarios need to be reported to the PM immediately so that matters can be resolved with the minimum of impact. For an ecologist, that may even be while you are on site and something hasn't gone to plan. For goodness sake don't get into your car and drive two hours back to the office to tell the manager that you weren't able to complete the survey. Get on the phone immediately and establish what can be done. OK, so you didn't have access to the site for the bird survey. Perhaps you could have covered the buffer at the far edge of the site for badger evidence in order to catch up or get ahead with another task. The main thing, when things are not going to plan, is that there needs to be quick and effective communication. If people are scared or reluctant to discuss such matters then this only puts the project and everyone associated with it at risk. In normal circumstances the only thing that someone should be hit hard on is for not communicating bad news quickly enough. Create the right culture within the project management team and you will reduce risks substantially.

Using the Gantt chart approach, on any given day you can draw a line down through all of the tasks and milestones. Everything that has been filled in up to or beyond that line is either on or ahead of target. Everything that has not reached that line is behind target. To demonstrate this, within Figure 7.6 I have inserted such a line (in bold) just after 31 March, and you can see that at this point in time one of the targets (client updates) has been missed.

If things are going to plan then that is good news, but seldom would everything go precisely as planned. Some tasks might take longer to complete, while others might be completed ahead of schedule. Whichever of these two scenarios is taking place you should be constantly revisiting the plan and making the necessary amendments. These amendments should always then be worked fully through as to their impacts on all of the other tasks that still need to be delivered. If something is behind plan, and is now going to be completed later than initially envisaged, how does that impact upon everything else? Likewise, if something is ahead of plan, does that now have the positive knock-on effect of enabling you to move other tasks forward? You should always grasp such opportunities.

Bearing in mind the customer, the other stakeholders and those identified as being involved or included at a task level (all as per your RACI allocations), any changes being made to the plan need to be communicated to everyone who needs to know. It is crucial to keep all of the key players at task level in the loop and informed as to what is going on. Think about the issues that could arise through poor or mistimed or no communication in this respect.

Table 7.6. Traffic light reporting

Component	Deadline	Red Amber Green	Required action if Red or Amber
Field surveys	28 June	Green	No comment required
Reporting	31 July	If Red or Amber	Include detail as to what is required to rectify matters. This would include who is responsible for completion of the required action and by when. In addition, the impact upon the plan, allowing for resolution, would be documented, as well as the risks to the plan if the matter isn't resolved.
Invoicing	14 June 31 July	Green	No comment required

Having now established that progress to plan needs to be monitored and people need to be kept informed, at what point and in what manner do you do this? First of all, if something has changed, and that change is going to have an immediate impact somewhere, you must straight away let everyone concerned know what has changed and how it needs to be reacted to. Aside from that, within the project plan itself, there will often be review meetings and progress reporting points incorporated. These reviews may involve informal discussions, meetings, email updates, formal reports, or indeed a combination of all of these things. In addition to predetermined review dates (e.g. monthly), other review points may kick in as a result of key achievements being arrived at, irrespective of when these occur.

The items that should be considered in progress reporting may be quite wide-ranging, and obviously would include each of the component areas identified, perhaps even to individual task level where necessary (e.g. when one particular task is putting a component at risk). Against each of the components and headings there would be a mechanism for showing whether or not things are going to plan.

Let's now consider informal updates that sit outside of the normal review schedule. My advice here is that they should always be followed up in writing, ensuring that everyone who needs to know is copied in. And it is not only what is happening next, or where you are ahead or behind plan, that needs to be documented. It is also important to tell people when specific tasks have been completed. This removes any uncertainty that someone external to the project management team may have as to whether or not something has actually been done. So, when something has been completed, tell everyone that needs to know.

One common approach uses the analogy of traffic lights, where 'green' denotes that everything is on plan and 'red' means that matters are behind plan, or are tracked to be behind plan until the reason for the blockage is rectified. 'Amber' signifies that there is a risk that the said item may fall behind plan unless corrective action is taken to put things back on track. Accordingly, in such circumstances, action should be taken immediately in order to prevent this from happening. In theory, if you deal with all of your 'ambers' effectively then the risk of a 'red' occurring is far lower. To demonstrate this approach, Table 7.6 uses part of our ongoing example.

In addition to the project-specific components, there would also be other areas where progress may need to be reported upon. Examples of generic items that may appear across larger projects are financial performance, quality of outputs, health and safety, complaints received, risk management and public relations activity.

As well as the production of formal updates for the project management team, the customer and other stakeholders, there may very well be, on occasions, opportunities to keep wider audiences abreast of what's going on. In such scenarios the production of newsletters or announcements via social media may be considered beneficial. A website could even be produced specifically for the project, if it was deemed cost-effective to do so. The reason I mention this area specifically is to make you aware that at a higher level, for larger projects, these methods may be used and as an ecologist you may very well become part of a 'good news' natural environment story that is being reported upon to the wider public.

CLOSE

As the project draws to a conclusion it is important to begin to consider the 'close' part of the overall process. This needs to be taken account of before the project actually finishes, as there may be matters that need to be discussed while the project team is still in place. It's easier to call upon this resource when the project is still live and people are still where you expect them to be. Also, things are still fresh in everyone's mind at this point, whereas weeks from now, once people have moved on to other projects, it is frightening how quickly they will forget the detail.

The first area to review, quite simply, would be: Has the project plan achieved the objective as required, to the right standard, on time and within budget? There may very well be benefits to documenting and communicating throughout the project management team, the customer's team and all of the stakeholders what has actually been achieved. When delivering such communications at the end of the process it's important not to lose sight of the methods you have been using to communicate up to now, and the audiences you have been engaging with (see *Monitoring progress*, above).

It is possible that a formal report documenting what has been delivered is required. If this is the case, are you able to produce this and copy in all stakeholders? In doing so, it is extremely important that you fully and properly acknowledge and thank all of those who have been involved with the successful result. This needs to take account not only of those within your business but also those working externally (e.g. in the customer's and/or a supplier's business). Too many times I have seen reports and presentations about projects that make no mention of, or quickly gloss over at the end, the appropriate acknowledgements and thanks. Rather than have it at the end, it should be right up there at the beginning. As PM, what you have overseen and delivered would not have happened without the support of others. You are reporting on behalf of lots of people who collectively delivered.

While you are working towards the close of the project you must also ensure that everything that was supposed to happen has actually happened. I am talking now about the various tasks, commitments and promises made that may be overlooked when the spotlight is on the overall objective. Even though they do not appear to be hugely important in the broad scheme of things, completion of all these smaller things is a reflection of your integrity, credibility and professionalism. Also, to you these things may not be that important, but to others they may be extremely so. The sort of areas that need to be checked are as follows:

- Does everyone who may need to find information in the future know who to approach or where to find it?
- Is there anything in the original agreement, or that has been promised since, that needs to be concluded? A review of all of the correspondence on file would be wise.
- Have all emails and the like been responded to?
- Have all suppliers been paid?
- Have all invoices been issued and settled?

REVIEW

While in the 'close' phase of the project it is wise to seek feedback from everyone who has been involved. This is an ideal opportunity to establish what has gone well in their eyes, and most importantly what could have been done better. Remember to use good feedback technique in requesting, acknowledging and handling the information that comes back to you (see Appendix 1, *Feedback: get rich quick*). All of this is going to make your handling of the next project even better.

It may be that the best way to carry out a review would be by email or telephone with those who were involved. Depending upon the logistics at the end of the project, however, a review meeting might be considerably more productive, as people can immediately hear each other's thoughts and debate matters in order to reach considered, collective conclusions. When holding such a meeting you should think carefully about who should be invited. Do you wish to include the customer and/or other key stakeholders, or should you involve only those from your internal project management team? Perhaps separate meetings with different groups of participants may be the answer? It's difficult to be prescriptive here, as in different situations, different approaches may very well be warranted or need to be considered.

Many projects are not big enough to warrant such an approach to the review phase, or it may appear a little 'overkill', or it may just not have been thought about. Also, there are often so many other things going on that finding the time to discuss yesterday's concluded work is going to be difficult, if not impossible. The thing to always remember, however, is that unless you recognise where your performance can improve, how are you going to make any improvements? It is just such a vital part of the circle of positive evolution from a self-development perspective (Figure 7.7), as well as from a project management perspective, that I cannot stress enough how important it is to carry out reviews of all projects, whatever their size or complexity.

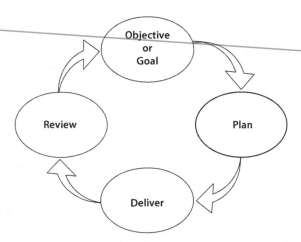

Figure 7.7 Development cycle (project management teams and individuals)

Chapter 8
REPORTING

I was working on the proof ... all morning, and took out a comma. In the afternoon I put it back in again.
Oscar Wilde (1854–1900)
Irish playwright, author and poet

Reporting, in various guises, makes up a substantial amount of the ecologist's time relative to their office-related tasks, and even more so when you include the other work (e.g. analysis of data and mapping) that also needs to be completed ahead of a report being ready to be issued to a customer.

In practice, in different ecological consultancies, there is a wide range of acceptable approaches and styles, depending upon the purpose behind the report. Nonetheless, it would make a lot of sense to adopt a common approach across the sector, and the high standards expected in certain circumstances should be held up as the norm across all reporting irrespective of the subject matter, the objective or the purpose of the report. In order to encourage its members to adopt a common high standard in their reports, CIEEM (the Chartered Institute of Ecology and Environmental Management) has produced a guidance note for report writing (CIEEM 2015), as well as specific guidelines relating to ecological impact assessment (IEEM 2006) and preliminary ecological appraisal (CIEEM 2013). Needless to say, these documents should be referred to and followed when writing reports. Furthermore, for certain subjects (such as bats) more specific guidance is also available (e.g. Hundt 2012).

In this chapter I am going to put forward ideas which are intended to complement the detail available from other sources. I am going to focus more on the techniques for producing reports than on the content, albeit content will need to be discussed to a point in order to cover the subject more fully. In some instances, the method I describe is not necessarily the only way to carry out reporting correctly. As with many things in the business environment, there are lots of ways to get it right, and also lots of ways to get it wrong. What follows will hopefully help you to avoid many of the pitfalls that could impact upon your performance as a report writer.

The bulk of people within the ecology sector, understandably, like to be outdoors in amongst the habitat and creatures they are so passionate about. Being in the office just doesn't hold the same level of enjoyment, and being metaphorically padlocked to a desk in order to get that report completed is just horrible. Yes, many of us do not enjoy reporting. Not necessarily because we don't like writing or producing something that documents all of our efforts, but more because almost invariably when a report is required this means that it needs to be done by a certain date or time. The dreaded 'D' word is lurking in the corner, waiting to pounce should you fail to hit the target. Deadlines: don't you just love them?

And that's not even the whole picture. As well as the deadline, the quality of what is produced is of huge importance. The report is required for a specific reason and is the shop window for all your efforts. The customer isn't really that interested to hear about how many weeks you have sat in the freezing rain doing the vantage-point bird surveys or how many dawn bat surveys you have carried out this month. All that really matters, from the customer's perspective, is that the report is produced on time and is fit for purpose. Having achieved this, the customer is usually only interested in three things, as described to me by Reuben Singleton of Tweed Ecology Ltd. Firstly, is their project going to be delayed as a result of your findings? Next, are there any timing constraints that will impact upon or apply during the construction phase of the development? And finally, are there any additional costs arising over and above what they were already allowing for?

WHAT IS REQUIRED?

It looks like we are back to one of those *'What's the job?'* scenarios (see Chapter 2). OK, before you can begin typing away at any report there are some fairly fundamental questions that need to be answered, these being as follows: What were you asked to do and why was it needed in the first place? What does the customer require and for what purpose? To whom (i.e. the target audience) is the report aimed? Who else, at whatever stage or whatever level, may also need to refer to the report? Broadly speaking, all of these questions can be considered under two higher-level headings:

1. The objective and purpose
2. The audience

The objective and purpose

At the very end of the process, once the report has been completed, the question that will be asked is: Will this report satisfy the objective? It therefore makes sense to not wait until then to ask this question. Instead, challenge what you propose to produce at the front end of the process, before you start using time, energy and resources, and potentially going off in the wrong direction.

What, precisely, was the question your customer asked you to answer when the work was commissioned? Surely this would always be obvious, but it has been known for someone to start working on a report with the assumption that the objective is the same as in every other similar case they have dealt with in the past. Then, later on in the process, it becomes very evident that they have been spending a lot of time and effort addressing questions that are beyond the brief, or, conversely, failing to address the brief. You must, therefore, be absolutely certain before you start that you do fully understand the objective. An example of an objective might be, '... *to report upon the findings of a breeding bird survey at the site where the client is planning to build houses'.* That all seems pretty straightforward, doesn't it? So let's get typing. Or rather, let's not – not just yet. There is something else that needs to be considered. That is, for what purpose is the report going to be used?

Depending upon that purpose, there may very well be a specific approach, writing style and/or template that will need to be used. For example, it could be purely to provide the results of an ecological survey (maybe a breeding bird survey or an otter survey) where one particular approach would be appropriate. On the other hand, if you were being asked to produce an Ecological Impact Assessment (EcIA) for a chapter within an Environmental Statement, a very different approach, style and structure would be required.

In order to recognise right from the start what is required, you should consider fully the objective of the work being carried out along with the purpose for which it is required (e.g. ecological survey to be submitted as part of a planning application). To assist you with this the CIEEM guidance document already referred to (CIEEM 2015) summarises the different types of reports that are typically required, along with their specific purposes.

The audience

Your audience usually falls broadly into three groups. Firstly there is the customer, then there are those who have required the report to be produced (e.g. a planning department), and finally, there is the wider audience, consisting of those who at some stage, for whatever reason, may also need to refer to the document (e.g. an ECoW during the construction phase trying to establish precisely where that otter holt was).

Your customer's opinion on the report being produced is extremely important. After all, they are the ones who are paying for it and they need to know that it will achieve its purpose. Quite often there will be customers who do not fully understand what is required in this respect and they will, very much, be guided by you as to what is appropriate for the given circumstances. On other occasions a customer will know precisely what they require from you, as well as the format in which it should be presented. A good example of the latter would be when you are subcontracted by another ecological consultancy for a specific piece of work. Either way, it makes sense for someone in your business to have a conversation with the client in order to fully establish exactly what it is that you are going to produce at the end of the day. As well as this being professional, it will potentially save time. Customers may have their own templates or styles that they want you to use, and/or they may want to see a draft report ahead of them signing off the final document. If you don't ask, you won't find out soon enough. Alternatively, you may find that someone in your business has previously produced a similar report for the same customer, and substantial feedback was forthcoming at that time as to that customer's preferred approach and style. Imagine the frustration – both for you and for the customer – in having to go through that whole painful process again.

The writing style used within the report must wholly consider those that the report is primarily intended for (i.e. whoever the customer is going to send the report to). Is there a style that they would be more comfortable with, or indeed expect to see? For example, a report aimed at the academic world would be written very differently to one aimed at a local authority planning department. Different audiences, depending upon the type of report and its purpose, may also require different reporting structures or indeed specific templates to be used. Bearing this in mind, rather than cracking on and assuming that you know what you are doing, ask the questions of the right people at the very start.

From the target audience's point of view, what style are they accustomed to seeing? What style would they prefer to see? You may think that the answer to both of these questions would be the same, but this is not necessarily the case. Sometimes, because of our scientific backgrounds, we ecologists can lose sight of the fact that many of the people who are required to see our reports are not that knowledgeable about our pet subjects, or not of a similar background, or, if they are, their expertise lies in completely different fields. Day after day they go through the pain of trying to make sense of all of these reports crossing their desk, in order to come to appropriate decisions. But because this is what they always see, it is not necessarily what they would prefer to see. In many instances most people just want things to be as easy to understand and as accessible as possible. Don't be afraid to just put it all down in plain English, unless of course there is genuinely a good reason why it has to come across in a way that demonstrates your all-encompassing sumptuousness of acquaintance with the unveiled characteristics of homeothermy and heterothermy in *Myotis daubentonii*. Find me a rocket, I'm moving planet!

On the one hand, yes, your report needs to be written to a high standard and contain all of the technical detail required. However, this does not mean that the language you use to describe your findings and conclusions needs to be aimed at the highest level of intellect. In fact you should adopt the complete opposite approach, in that the report should be able to be understood by anyone, irrespective of their ecological pedigree or academic achievements. As my good friend David Darrell-Lambert (Bird Brain UK Ltd) has said to people on many an occasion, *'Does my granny understand what is being said?'* If David's granny can understand what you are talking about, then you are probably on the right tracks. By the way, his granny was never an ecologist (or at least he has never mentioned that she was). In effect, what this approach achieves is that not only can your primary audience understand the document, but so can anyone else who may need to do so at some stage in the future. Accordingly, avoid scientific jargon and too many statements that only an academic can understand. Otherwise you are potentially restricting the effectiveness of interpretation and the practical application by others in the future. Surely you would want all of your hard work to be of the greatest use possible. Aim to produce something that is targeted to the audience required, but at the same time is accessible to the masses and not cloaked for the few.

REPORT STRUCTURE

Across many sectors in industry a fairly standardised approach to reporting exists, in which a report is arranged in three main parts (Figure 8.1). Generally speaking it is good to think about how these three parts may relate to different people, at different levels, taking in the information you are putting across. As you travel down through the levels the amount of detail increases, but the message remains consistent right from the very start.

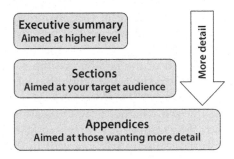

Figure 8.1 Report structure

The executive summary

The first part of the report is commonly referred to as the executive summary. This should be viewed as a high-level summary of the contents. Anyone who reads the executive summary and then goes on to read the rest of the report should encounter no surprises as to its content. The purpose of an executive summary is to enable someone at a high level within an organisation to quickly appreciate the content of the entire report without having to read the rest of it. With all of that in mind, and allowing for the fact that some people will only read the executive summary, you have to be very careful that it fulfils its function. When it comes to writing this part of a report, you need to ensure that you cover everything that a senior manager, for example, needs to be aware of.

Ideally, the most effective way of writing an executive summary is to complete it after you have finished the rest of the report. If for some reason you started working on it earlier (perhaps because you felt it was a fairly straightforward message), you most definitely need to revisit what you have written once the whole report is complete, to ensure that it is still an accurate reflection of what the report in its entirety is saying.

For ecological surveys you would generally expect the executive summary to say, at a high level, what the objective was, for what purpose the report has been written, what work was carried out, what the results were, and how these findings impact upon the customer's proposed actions. In respect of the last of these, it also needs to say what must happen next. Finally, a summarised conclusion about the report's findings should also be given.

The sections

This part of the report contains the main body of what you are wishing to say, and should very much be targeted towards the audience the report is primarily aimed at. Within our sector you would often find that it follows a fairly typical structure, section by section. Occasionally, some amendments to the standard structure can be made, depending on the subject matter and purpose of the report.

In the section on *Report content* (see later in this chapter) I will specifically cover the individual sections that typically occur. For now, I merely wish to make a point about how you should view the writing of sections, in comparison to how you would approach the executive summary and appendices. The main thing to remember, and the point I wish to make, is that someone should be able to make fairly good progress through all of the sections within a report without getting bogged down in detail or distracted by information that takes them away from the key messages you wish to give them. Therefore, while you are writing the sections you must constantly ask yourself whether or not you are guilty of providing too much information or distracting from your key messages. If you are, then the place to put all of this extra stuff, assuming it is relevant in the first place, is in the appendices.

The appendices

The appendices are where you put all of the more detailed information along with maps, plans, additional supporting pictures and the like, all of which, if left within the main body of the report (i.e. the sections), would bog down the reader or cause unnecessary distraction. Although this is where you would include detailed supporting information (e.g. raw survey data), you should still resist overloading the report with information that is likely to be beyond the interest of a reader or doesn't give any added value. If there are huge amounts of raw data you should seriously consider summarising it into a manageable length. If the customer requires to see all of it in its raw state then this can always be catered for separately, in a suitable format.

I sometimes affectionately refer to this area of an ecological report as the 'geek pages'. This is where people like me (it is possible that some people might consider me a bit of a geek) go in order to get the pieces of additional information they are really interested in poring over. Accordingly, this is also where you will find lots of information that many people involved in the process don't really care that much about, and some readers will therefore spend little or no time in this part of the report. Considering this, it is crucial that there is nothing contained within the appendices that would impact upon anyone's interpretation based on reading purely the executive summary or the individual sections. Otherwise there is a high risk it will not be picked up.

REPORT CONTENT

In order to discuss the report content I am going to use a generic reporting structure. The main headings that will be covered are as set out in Figure 8.2, which in effect are those that are usually included, in some shape or form, in an ecological survey report. As discussed earlier in this chapter, the CIEEM guidance should be referred to in conjunction with many of the points I am making here. My intention is to complement the guidance, as opposed to contradicting it. If anything described here does unintentionally contradict the guidance, or if you are ever in any doubt regarding reporting rules and approaches, please do follow the guidance applicable at the time. Within this section, and also continuing into the section on *Practical advice*, below, I am going to provide as much advice as I can, in the way of helpful hints and tips on preparing and completing reports.

Title page
Contact details
Table of contents

Executive summary

Sections:
Section 1 - Introduction
Section 2 - Methods
Section 3 - Results
Section 4 - Discussion
Section 5 - Conclusions
Section 6 - Recommendations
Section 7 - References

Appendices

Figure 8.2 Report structure (an example)

Title page

The title page appears at the front of the report (where it may double as the front cover) and within this area of the document there are a number of points that should ideally always be shown. Ask yourself what you would want to know when just looking at the front page of a report. In Figure 8.3 I have included the points that you should seriously consider having there. It is surprising how often the most basic of information isn't present. For example, it is not uncommon to see ecological reports where fairly basic information such as the address or OS grid reference of the site is not provided anywhere, let alone on the title page.

Although not essential, quite often a suitable site picture, as shown in Figure 8.3, does no harm in letting the reader identify with the subject matter (e.g. the site) of

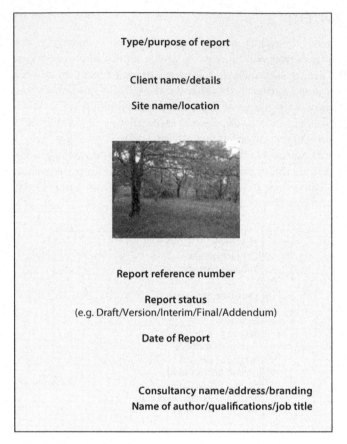

Type/purpose of report

Client name/details

Site name/location

Report reference number

Report status
(e.g. Draft/Version/Interim/Final/Addendum)

Date of Report

Consultancy name/address/branding
Name of author/qualifications/job title

Figure 8.3 Example of a title page layout

the report. If you are using a picture, then please do ensure it is typical of the site or the features that the report is discussing. You may find it hard to do this with a single picture, in which case you could consider a group of photographs that create the right impression. You should also remember that a photograph can only be used if you have the right to use it (in other words, if it is your own, if you have permission from the photographer, or if it has come from a copyright-free source).

Contact details

It is important that one of the first pieces of information provided within the report tells the reader who to get in touch with should they have any queries about the report contents. Accordingly, the person's name, qualifications, job title, email address and who they work for should be the minimum that is shown. In addition, their business telephone number would also be useful.

It strikes me as odd when this type of information is not easily accessible. Imagine the unnecessary hassle that someone in your customer's office or the planning department might have to go through in order to find out how to get in contact with someone about the work. To not show this information is in effect saying that you do not want to be easily contacted about matters relating to the report. A service-sector business that adopts this veiled approach is not in any way helping to demonstrate a good service ethos.

Table of contents

A table of contents should appear at the beginning of the report. You can debate precisely where it should appear, but either immediately before or immediately after the executive summary would be acceptable. Although in a short report the table of contents arguably doesn't add much, it is something that you would always expect to see in a professionally produced document. It is therefore desirable to include one. For much longer or more complex reports it fulfils its main purpose of enabling the reader to quickly navigate to areas of the report that are of specific interest. In today's digital world the table of contents can be further enhanced by software features that allow the user to click on a page or section number in order to be taken directly to that page.

Executive summary

The executive summary, the first main component of the report (see Figure 8.1), has been discussed in some detail in the section on *Report structure* earlier in this chapter.

Sections

The part of the report containing the sections that make up the main body (see Figure 8.1) has been discussed in general terms under *Report structure*. I will now proceed to discuss each of the sections that you would expect to see in many ecological reports, and I aim to provide some useful points for you to consider when preparing these.

Section 1 – Introduction

This is one section of the report that you can start working on almost immediately after it has been given to you. A lot of the information you require in order to complete many parts of the introduction will be known at the outset when the work is commissioned.

Within the introduction you would want to ensure that the information that a reader would find useful, before describing specifically what you did, is available. Therefore, it's important here to describe the objective of the work, the purpose of the report, and also your client's proposed intentions (e.g. to develop the site into a housing estate). Describing the site as it currently exists would also be of great relevance (e.g. derelict farm buildings, or coniferous plantation). Also, if your client does not own or manage the site you should say who does, and if your client does own or manage the site then state that this is the case.

Details regarding the location of the site (address, post code, OS grid reference) are essential here. You and your client know exactly where the site is, but a reader may very well not. Make it easy for them to find the specific location you have surveyed. It is often useful to refer to a generic site plan of the area in question, within which you could incorporate the development and survey boundaries, along with any applicable buffer zones. Do not, however, include the plan itself within this section. Instead, make reference to the plan and point the reader to the appendix within which they will find it.

Headline information about the work you carried out at the site and the period of time over which this was done would normally be shown. It would also usually be essential to say for how long the information you have provided within the document can be relied upon (in other words, its shelf life). Do not, however, allow yourself to put in detail regarding the results, discussion and recommendations. Remember that this area of the report is an introduction, establishing the context and preparing the reader for what is about to follow. It is not intended to be a summary of everything that has occurred, although you may very well wish to refer to other parts of the report, as well as to a few key references, such as ecological reports previously carried out at the same site.

In some instances the introduction might include contact details. This would be acceptable if it hasn't already been catered for elsewhere in the report, as discussed earlier in this chapter.

Within the introduction you may feel that some of the detail is repeating information contained elsewhere (such as within the executive summary or on the title page), but don't be influenced by this. The executive summary is actually repeating some of what's in the introduction (not vice versa), and this section, the function of which is to introduce the report, needs to stand on its own two feet. As for the title page, that is there to head up the document, and in doing so it should make it easy for someone to see certain basic details without having to search for them. It covers some of the information contained within the introduction, but for a different purpose.

Section 2 – Methods

Like the introduction, this is also a section that you can start working on almost immediately. It should be very clear, even ahead of the field work commencing, which methods you are going to use and whether or not adaptations are likely to be needed. It would indeed be a concern if this wasn't the case. Yes, there may very well be alterations to what was initially intended along the way, but getting the bulk of the information into the report early on will save time later, and it should be fairly easy to edit as and when necessary.

Within this section you are focusing on how you conducted the work and why it was done this way. In the process of doing this you would almost always be referencing approved methods and saying when and why you have deviated from these standard approaches. Don't be afraid to say that you haven't fully followed a recognised methods if that is the case. It would be wrong to give the impression that a method has been fully used when it hasn't. Hopefully, if a method has been adapted there have been good reasons why this has been the case and therefore you should explain or justify your rationale for deviating from the norm.

In laying out the methods section it is usually useful to consider it in two parts, assuming both have occurred, of course. Firstly, the desk study, and secondly, the field work.

As the desk study usually helps inform matters ahead of field work, it is preferable that it is carried out and discussed before any field surveys. If a desk study was carried out, then describe the process for gathering the information, but not the results; those go into the results section. You would also say who was approached for data and/or what other resources were referred to. If on the other hand a desk study was not carried out, it would be useful to say so, so as not to confuse a reader into thinking that it has been mistakenly omitted.

You would then go on to describe the field surveys that were carried out. Specifically, for each survey event, you would describe which part of the site was covered on that visit, when the survey took place and who did the work. When discussing the survey team you should remember to provide details of their relevant experience, qualifications, licences, and so on. Details of specialist equipment used during the field work along with particular reference material and identification keys used should also be given. In some instances, if appropriate, analysis and statistical software packages used should be detailed. Finally, any constraints (e.g. restrictions in access) and limitations (e.g. caused by weather or seasonality) should be catered for within this section.

Section 3 – Results

This area of the report cannot really be worked upon that much until you have the results, albeit there are aspects of preparation in anticipation of the results that could be started in the early days of a project's life. Once results start to come in, however, you should be progressively working upon the analysis of data, mapping, and production of neat copies of survey forms. It makes no sense to delay all of this work until the end.

As with the methods section, it is often useful to consider the results section referring, separately, first of all to the results of the desk study (if one was carried out) and then to the field survey results. When working on this part of the report, always ensure that what you are producing relates to factual information about what has been discovered at the site. The *'what does it all mean?'* bit will be dealt with under the discussion, conclusions and recommendations sections. I find it useful to think about the results against the backdrop of *'say what you saw'*.

In relation to the desk study results, provide a summary of the information that is relevant to the site and the objective of the report. There may very well have been considerably more detail provided by the records centre, and if so you can always elect to put all or some of that additional detail into the appendices. In fact the results section will often contain a number of cross-references to far more detailed information that is placed in the appendices, especially when it comes to details of the field surveys.

When describing the field survey results, ensure that you don't discuss the methods adopted. That should have already been covered in the methodology section. Similarly, resist interpreting or talking about what the results might mean, as this will be covered elsewhere, in the discussion and conclusions sections. Simply describe what was found, where it was found and when it was found. Keep the data at a reasonably high level, but ensure that everything relevant is mentioned. Do not bog the reader down with detail that they don't need to know about in order to work out what's happening. Items such as field maps, spreadsheets full of data and survey forms should all go into the appendices.

There may of course be occasions when the level of detail provided is restricted for other reasons: you will, for example, not show the precise locations of badger setts or raptor nests. Some of the information we report upon may end up in the public domain (such as on a local authority planning portal) and it could, sadly, be accessed by those intent on criminal activities. The approach to adopt in such circumstances would be one of two routes, or possibly a combination of both. First of all, you can take that particular report you are concerned about and make it 'confidential and unsuitable for being placed on a public domain'. This can usually be achieved by stating so on the title page and alerting the planning authority as to why there is a need to be extra-careful on this occasion. Alternatively, you can give more general information about specific locations of sensitive data within the report and provide the more detailed information, if required, separately.

Section 4 – Discussion

This part of the report cannot be fully tackled until all of the results are in. At the last minute of the final survey something can crop up which changes everything. This is, in part, why it's important to manage your customer's expectations while you are still carrying out a survey programme. It is very tempting to tell them that everything is going to be alright, build up their hopes, and then have to give them contradictory news later on.

When it comes to completing the discussion section, as the survey programme progresses you will be invariably thinking about the impact of the results (positive or negative) on your customer's aspirations. As thoughts come to mind place them into this part of the report as bullet points and snippets of ideas. It is also a good idea for you to have a bit of a brainstorm (see Appendix 3, *Effective brainstorming*) with colleagues about the project and the implications surrounding what is being found or not found at the site. As you begin to develop the discussion, taking time to properly think out your thoughts is certainly worthwhile, remembering to substantiate these, when appropriate, with supporting referenced material.

The discussion section is essentially taking the *'what you found'* information from the results section and moving it on to discuss *'what it all means'* in the context of the objective of the project, the purpose of the report and the proposed work that is planned for the site. When properly starting on this section you should definitely reacquaint yourself with the file history and be confident that what you are going to spend your time discussing are the points that are relevant to your customer's plans. So, you need to ensure you are spending enough time talking about the right stuff and not devoting space to what are, relatively speaking, distractions. Keep it relevant.

Section 5 – Conclusions

The conclusions section allows you to reinforce the main message being given by the report overall. Therefore, it provides a summary of the report and arrives at an overall conclusion that takes account of everything that has been found. It will then take all of this information and relate it to the potential impacts of what is being proposed by the customer. As this section is driven by what has been already found and discussed elsewhere within the report, no new information should be introduced at this stage. It should only take account of what has already been presented elsewhere. If you do find yourself dipping into new territory you should go back to the appropriate section and add in the relevant information accordingly.

This section should cover how, and to what degree, the report as a whole meets the objective it sets out to achieve. The conclusion should provide a general review of the results, relating this to the points developed within the discussion section, and then arrive at what has been concluded overall in relation to the proposed activities. To quote the CIEEM guidance (CIEEM 2015):

> Where a report has been prepared to support a planning application (e.g. an EcIA), CIEEM would expect the report to provide a clear statement in the conclusions as to what the likely outcomes are for biodiversity if the proposed development is granted planning permission.

Having arrived at a conclusion, the section should take the likely outcomes and demonstrate, assuming it is the case, the action that is required in order, for example, to mitigate and/or compensate for the proposed activities or changes occurring at the site. The action suggested is then documented more fully within the recommendations section. Relating to this, in many reporting structures you find that the conclusions section appears immediately before the recommendations section. The thinking here being that the conclusions section is, in part, setting up the rationale behind the more detailed recommendations that are about to follow.

In some working environments certain people may read through the conclusions section and the executive summary, ahead of, or instead of, reading the entire report. Therefore, you should ensure, in a similar way to how you have approached the executive summary, that this section accurately conveys the message that the report is making overall.

Section 6 – Recommendations

The recommendations section is the tipping point where what has been discovered, discussed and concluded crosses over to what course of action, if any, needs to take place. Although it is often called recommendations, it can also refer to requirements. As such, in order to prevent confusion or misinterpretation by a reader, entitling the section *Recommendations and requirements* may be appropriate, depending upon the purpose of the report and, of course, what you might be putting forward as recommended or required actions.

When presenting recommendations and/or requirements within a report it is wise to have a discussion with the client about what you are proposing to say in the context of what they are seeking to achieve. It is also important to separate in your mind what is a recommendation (i.e. something that *should* happen) and what is a requirement (i.e. a much stronger message: something that *must* happen). Be very clear as to how you word all of this. By way of example, saying that something could or should happen may not necessarily be interpreted by those concerned as meaning that it *has* to happen, for the words 'could' or 'should' may be interpreted as giving those concerned an option. If they have to do it, say so in language that cannot be misconstrued.

This section also needs to be worded carefully because it is allocating tasks to people beyond your own business. A good approach is to use an action plan template which lays out very clearly in chronological order what needs to happen, who is responsible for making it happen, and when it needs to be done by (Table 8.1). Generally speaking this approach is far easier to follow than simply setting out the requirements in the text, even if there are just a handful of action points that need to be described.

If there are any concerns about the proposed works at the site conflicting with ecological features, then you would usually wish to ensure that something has, at least, been recommended within this section. In such circumstances there are certain points that should always be considered as standard, including the provision for toolbox talks (TBT) being given to site contractors. Incorporated within these talks would be the identification, on the ground, of any areas where constraints apply. At this same point in time it makes complete sense to physically mark out any buffer zones and the like. Following on from this would be all of the additional actions that are recommended or required.

There is no reason why pictures, maps, plans or diagrams cannot be referred to or incorporated within this section, or within the action plan itself. Doing this removes any doubt as to the location or meaning of any of the proposed action points. It's always worth bearing in mind that someone (such as an ECoW) may, at some stage in the future, be looking at the report with a particular focus on this section. You should therefore ensure that the action points are cross-referenced to

Table 8.1 Action plan template (with an example, AP1, included)

Reference number	Action to be taken	Who Is responsible?	Deadline
Give each action a reference number so that they can be referred to more easily	State precisely what is required and ensure that the words used describe exactly the instruction you intend to give and that there is no ambiguity	State who is responsible for ensuring that each individual action point is delivered	Give the deadline or a timeframe within which the action point must be satisfied
AP1	It is a requirement that prior to the commencement of any contract-related works within the site boundary that a toolbox talk (TBT), given by a suitably experienced ecologist, must be delivered to all site contractors and associated operatives who may at any time be working within the site boundary	B Jones Owner of Site Contractor Ltd	Prior to any works occurring within the site

where the relevant supporting information that may prove useful on the ground can be found.

If your results are inconclusive, or further surveys are required, this section of the report is where you would formally document precisely what is now required in that respect. Finally, the point beyond which the current results reported upon are no longer reliable and further surveys need to be commissioned would also be repeated here, this being in addition to the same message being given within the introduction. It is wise to repeat it within the recommendations section just in case someone is focusing purely on this area of the report at some point in the future. For instance, you might include an action point along the lines of, '*If construction work does not commence prior to 1st April then further surveys at the site relating to bat roosting activity within Structure A will need to be carried out in accordance with recognised guidelines, by suitably experienced ecologists.*'

Section 7 – References

Referencing should follow a recognised approach throughout the report, with all cited references brought together within this section. The Harvard (author–date) system is the most widely used style in ecological reports, and is in fact the manner in which references have been shown in this book.

One thing to really focus on when producing your report is to ensure that all of the references mentioned throughout the text also appear accurately cited within the references section. Having a system for checking this off as you go and then double-checking during the proofread is essential. Something that sometimes gets missed, however, is that the author gets so embroiled in checking that references appearing in the main body appear in this section that they forget to repeat the process backwards. By this I mean you should also check that all of the references listed do actually appear elsewhere in the report. Occasionally a reference may accidently get left in the references section from a previous version of a template or an earlier draft of the report and is now surplus to requirements. This being the case it should be removed.

Appendices

The appendices area of the report (see Figure 8.1) has been discussed in some detail under *Report structure*, earlier in this chapter.

PRACTICAL ADVICE

Where to start?

Irrespective of the objective and purpose of the report, or the target audience it is aimed at, there will be a number of things that will usually be 'a given' when it comes to producing reports within any particular ecological consultancy. Matters such as where and how the business branding appears on the report; how pages, sections, subsections and appendices are numbered; the font size for different heading types; the font style to be used – all of this, and much more, will have evolved into the standard 'in-house' approach agreed within that business. If the report is for a specific customer or being used for a particular case you may find that despite what your business would normally do, on this occasion different rules need to apply. There may be a template you are required to comply with by an external source. A good example of this is when the person coordinating the overall reporting process for an Environmental Impact Assessment (EIA) would require all documentation being submitted from the various supporting contributors to follow the same rules in structure, style and the like.

Usually, at least in established businesses, there will already be reports previously prepared for clients of a similar nature or with similar objectives/purposes. These can be referred to prior to preparing the report that is now on your desk. In fact you would probably expect there to be draft templates for most of the recurring themes within a particular consultancy, and these templates will then be used as a starting point in preparing the report in question. In doing this, however, don't assume that what was previously produced or currently available as a template cannot be improved upon, or will not need to be adapted for this scenario. In particular, one activity you must really be careful about relates to cutting and pasting material into the report you are currently working upon, from elsewhere. If, on the one hand, you are doing this from a report that has already been produced by the business you are currently working in, then there are no real issues, provided of course you ensure that what you have transferred over is correctly reworded to reflect the new case. It is so easy to inadvertently leave in names, locations, OS grid references and the like from the previous document. Be careful, so very careful. On the other hand, an approach that would be totally unacceptable, and potentially in breach of copyright laws, is the copying of material from another firm's report and pasting it into yours, passing it of as your own work. This is a very risky and unprofessional practice and should be avoided at all costs.

For a number of new ecologists, when they are given the responsibility to complete their first formal report they will most likely will be introduced to a template to work within. The instructions given could be along the lines of *'All you need to do is fill in the blanks the best you can, and once you have finished we can have a look over it together and discuss where improvements can be made.'* And so it continues, report after report, with the good intentions being that with every new report the author will make improvements. Quite often the person showing new people what to do may not be at a particularly senior level, or, even if they are, they may not have been given any training themselves as to how to be an effective coach. Now, it would be wrong to suggest that this is always the way. There are, in many instances, far more structured approaches taken when introducing people to reporting for the first time. I particularly liked the approach that Steve Jackson-Matthews described to me: his preference is to give the person an example of a similar but unrelated report to look over first, which then leads on to a discussion as to why it is structured and worded as it is. Having done this he works with the trainee in building a road map for the report which is about to be written. During this they would then discuss fully what was required for this particular case, making reference to a map or plan of the area being developed in order to help the ecologist visualise more clearly how everything relates to the site. Strong guidance and coaching throughout these early days of reporting helps considerably as an ecologist gains competence and confidence about what is required in such scenarios.

There are, therefore, lots of hints and tips that an experienced person can pass on to someone in the early days of their report-writing responsibilities. In addition, these same hints and tips may prove useful to others who are, after some time, still regularly failing to deliver reports to the correct standard, as well as to those who are themselves beginning to take on the responsibility for proofreading other people's work.

Give yourself time

The biggest problem that many people encounter is underestimating the time it takes to produce a high-quality report. When the deadline is months away, you don't have the drive to get it started. Months later, when all of a sudden it is due tomorrow, you now have the drive (well, if you don't by this point, you never will!) but not the time. As a result, it is all rather rushed and potentially inadequate as it makes an emergency landing on your line manager's desk for proofreading. If only you had started earlier. If only the deadline could be pushed back a few more days. At this point I would like to refer you back to Chapter 5 (*Organisational skills*), but there's no time. You have a report to write. Stop! Remember, you should always make time to be organised, and if you do you will benefit from the extra time you have engineered for yourself. If ever you need to create extra time, you certainly need it here.

The answer is to start writing the report as soon as the case lands on your desk. Ask the right people the right questions on day one. Decide which template you are going to use and start populating the template (which is now your working report document) with everything that you know right from the very outset (the client's name, their address, the site name, the species being surveyed, the various bits of standard text regarding conservation, legislation and methods). In the course of time, as further information comes in (e.g. local records for the desk study), then add it in, and as each survey visit concludes complete the analysis required and proceed with populating the results section. You are going to have to do all of this in any case, so you may as well do it while it is fresh in your mind. As the case develops and you begin to have ideas about the discussion, the conclusions and recommendations, bring them into the appropriate areas of the report. I am not saying that you should attempt to have these all formally prepared at this stage, rather that you should use the evolving document to insert your ideas. It's just as easy doing this as scribbling thoughts down on scraps of paper in the file, and it is considerably more reliable than taking a mental note and continually trying to remember them. Referring back to *Organising your environment* in Chapter 5, putting everything in its proper place at the right time means that you no longer need to be cluttered by it bouncing around in your head. You may also have some crazy ideas that most likely will not end up in the final document, but sometimes crazy ideas open the door to excellent solutions. So document all of your thoughts.

Everything in its place

The main thing is to be disciplined as you go along. A report does not need to be something that only gets started after everything else is done. I once knew someone who typed up all of their results as they went along into a Word document. Later, once they started on the actual report they would start copying and pasting data from one document (with a different font and paragraph spacing) into the report document. Why you wouldn't just type it immediately into a report template, where it was going to end up anyway, is beyond me. Why do it twice, when once will do? Do you really have the time to be doing it twice? In the same vein, coming into the office and writing up your notes using pen and paper, and then having to type them into the report, pretty much word for word, is just duplication of effort. Yes, on occasion there may a need to have neatened-up versions of field survey forms or maps to be produced for the file, and I would accept that. What I am talking about here is writing out something that you know will eventually be typed into the report. If it's one of those tasks that no one is going to complain about if you stop doing it, then STOP (see *Manage your priorities* in Chapter 5).

In the early days of your report, view the document as threads of thought and patches of material that will eventually be woven into larger complete sections. You build it up word by word, idea by idea, layer by layer. You bounce from one part of the report to the next as thoughts and information come your way, and eventually you bring it all together in its completed form, in order to present the finished article to the customer.

As you put the report together, consider how the use of tables, pictures, plans, diagrams and maps can all help the reader interpret what you are saying. When using these aids to interpretation, it is very important that on each occasion the reason for using them is borne in mind. We once worked on a large bridge project where a bat roost was described in another ecological consultant's report as being on the west wing wall. A picture of the whole bridge, including the wing wall, was shown. That was it. There was nothing else in the text or in the picture to help us or the client in locating where exactly the roost was. The wing wall was about 10 metres high and 5 metres wide. It had, at least, 15 potential bat roosting areas within it and we had to treat every one of those on the basis that it might be the location that was being described. All of this due to the absence of an arrow on a picture. I have lost count of the number of times I have encountered something similar. It seems strange that I am describing all of this and not providing you with a picture to demonstrate the point. So here goes: see Figure 8.4 for an example of the good use of a picture in a report.

Likewise, with the use of maps and plans. For example, if you are going to go to the trouble of using a plan of the site within the report (and I certainly suggest you always do), it is not that hard with today's technology to add in details such as surveyor positions, transect routes, the location of findings or constraints (e.g. a buffer zone around a badger sett) that may be applicable. It's all very well describing these things in the text, but it is much easier for the reader to understand it if you show them. Also, text alone in such circumstances can be cumbersome and not easily transferrable into practical activities, if for example someone needs to find the sett location in order to accurately install the buffer zone signage you have requested.

Figure 8.4 South side of the property showing bat roost access point (white arrow)

Having the plan in front of them, in the field, greatly speeds things up and increases confidence that they are definitely in the right place doing the right thing. What are the implications for that same person being lost or confused? Lost time, meaning expensive. More time on site tramping through woodland, meaning increased health and safety risk. The badger sett not being adequately protected, meaning ... I am sure you get the picture. Insert your own arrows.

Having discussed the main points relating to the preparation of reports, both in this section and in the sections on *Report structure* and *Report content*, I will conclude with some additional hints and tips, in Table 8.2.

Table 8.2 Practical hints and tips on report writing

Subject matter	Things to consider
Writing style	The whole report should be written impersonally in the third person. Use of 'we', 'us', 'I', 'me' should not occur.
Headers and footers	Use these areas of the document to ensure that all pages have a record of the report reference, date, title and page number. Page numbers will help if pages are extracted at some point, showing that they were part of a larger document.
Abbreviations, acronyms and symbols	The first time one is used in the report, expand upon it to let the reader know what it stands for or means. We may all know what an ECoW is or who SNH are, but many people don't.
	If there are many, then the inclusion of a glossary should be considered. If a glossary is used then make sure it appears or is referred to at the start of the report. It is irritating to painfully work your way through to the end, only to find that there is a glossary after all.
Pictures	Can say a thousand words and should be there to support the text. Overuse of pictures that do not describe anything new or give useful information should be avoided.
	Avoid images that are larger than necessary, or too many pictures, which can give the impression they are being used to bulk out the report.
	Crop pictures so that they are suitable for their message.
	Use contrast and brightness to make them more appealing.
	Compress them, to save memory space, before sending to client in final report.
Maps	Need to be large enough so that they can be used in a practical setting. Keys, symbols and the like should be large and clear.
	As well as using different colours to show different types of record, use different shapes (allow for the possibility that someone might take a greyscale photocopy).
	Always ensure they are from sources that allow you to use them for commercial purposes. A number of regularly used sources for maps on the internet are purely for personal use, and using these resources in a commercial report would be in breach of the terms and conditions. If licensed, then the licence number must be shown.

Table 8.2 cont.

Subject matter	Things to consider
Figures and tables	Ensure that they are clearly labelled and support what is being said. They should always be referenced within the text, and before they appear.
	Always consider this approach, as opposed to lots of text. It gives the reader a break, and is usually far easier to understand when used in the right way.
Clarity	Use words that say precisely what you mean, and avoid ambiguity.
	If words that suggest uncertainty or lack of clarity have to be used, then you should explain why a more definite conclusion or approach is not being given.
Acknowledging others	If others, external to your business, have assisted with the work or given advice it would not be ethical to omit this from the report.
	Suggesting that all of the work was done 'in house' when it wasn't is creating a false impression to the client and could be a source of huge embarrassment later on should they find out.

PROOFREADING

Preparing a report is a bit like doing an exam. It is going to be proofread (marked) and you don't want a poor score from either your line manager or the client. In the same way that you may tackle an exam, a good approach is to get the easy stuff quickly done and leave time at the end to properly cover the bits that you need to really think about. The point that is often missed, however, is that you should ensure enough time is left at the end of the process to allow for proofreading and making the necessary adjustments. Therefore, when a report is due to a customer by a certain date, you need to allow for the additional time beyond when you think the report is completed, so that someone can proofread it and get back to you with any amendments required. Once it has been amended, the report may then need to go back to the proofreader a second time before being signed off.

At the proofreading stage, think of all the potential readers of the report, at all levels and within all of the key stakeholders, present and future, potentially relying on the details contained within your report. Very few of these people will have an ecology background. As I have mentioned earlier, the language you use needs to be clear and accessible to everyone. In addition, the report needs to be accurate, professionally written and well presented. The final document is a reflection of your personal brand, as well as that of your employer.

The role of the author in the proofreading process

Once you have completed writing the report, you still have an opportunity to ensure it is as accurate as it can be before passing it over to someone else for the official proofread. Proofreading the report yourself sounds fairly straightforward, but in practice you are really up against it. As someone once said, 'Don't ask me to proofread *my own writing. I always end up seeing what I thought I wrote.'* It is so true. The last person on the planet who should be carrying out a final proofread is the original author. It is extremely difficult for you, as the originator of the words, to see those random typos and grammatical errors, and you are never going to take yourself to task over something that isn't quite clear enough. You will find it very hard to spot the mistakes, especially perhaps in the case of common words that the eye normally skips over. You see what what you meant to write, not what you actually wrote, and to you the meaning is perfectly clear. If you understand it, why would anyone else struggle to understand what you are saying? The point is, however, that it is all of those 'anyone elses' that you are writing the report for, not yourself. (Did you spot the error in this paragraph?)

Nonetheless you must at least make an effort to ensure that the report is as accurate as you can make it before you submit it to the official proofreader. To help improve the chances of finding mistakes it's worth bearing in mind that for many people a hard (i.e. paper) copy of the document seems to be better than continuing to read it on a PC screen. So print off a hard copy and take it somewhere quiet to read. You do not want any distractions. Always read through the entire document again, start to finish. Always run a spel check. Do this at the end of your personal proofread – but remember that it cannot be relied on to pick up everything. Many words may be spelled correctly but completely wrong in the context, four example their and there. Always compress pictures and plans, and always update automated fields such as the section numbers on the contents page. After you have completed all of this, *now* it is finally ready to go to the official proofreader.

The report should be perfect in your eyes before it goes to the proofreader, so that the proofreader's job is to identify just the last few amendments that are still required. Taking the approach that *'It's OK, the proofreader will pick up any errors'* is unprofessional. It is akin to saying, *'My time is more valuable than the proofreader's.'* The proofreader is likely to be a senior ecologist, or your manager. If they have to make lots of amendments, or write paragraphs for you, then they would be perfectly within their rights to just throw it back to you and tell you they don't want to see it again until it's accurately and fully completed.

If you are ever in that position I would hugely recommend that an apology is in order, along with taking practical steps to ensure it doesn't happen again. You need to look very closely at the mistakes that recur within the report, or over a number of reports taken collectively, and then take this on board. If you don't, then you are at best not being effective, and at worst you may be making your line manager feel that you are not only unproductive but also impacting upon their productivity. Also, if the proofreader is suggesting that something is not clear, change it. Don't argue about it, and don't hold onto the original words passionately. Be thankful that the proofreader has averted potential confusion within the audience.

The role of the proofreader

The official proofreader will probably be someone senior to you in the business. As already discussed, their time is valuable (probably more so than yours), and in addition they will probably also be responsible for proofreading the work of other team members.

The proofreader needs to check that the report satisfies its objectives. Proofreading is not just about the document itself, but also involves referring back to the file and ensuring that what has been produced is fit for purpose and does the job. The process should also hopefully find any typos that have crept in, as well as provide critical comment on how the language might be interpreted by a reader. The proofreader should also be able to give technical support and advice to the author, to improve areas of the report that may be inadequate. In some businesses the proofreading process is split into two separate jobs: firstly a technical review, and later, immediately before release to the customer, a grammatical check.

It would be normal, in fact beneficial, for the proofreader to use software facilities such as 'track changes' and 'new comment' within MS Word. This allows the author to see immediately where changes are needed, and also to make them quickly (e.g. 'accept change' in MS Word). This approach can also act as an effective coaching session for the author, whereby they are able to learn from their mistakes and become better acquainted with the style that a more experienced person likes to adopt in reports. Hopefully, this also helps to ensure that the next report is at a better standard, and thus saves the proofreader time in the future. If you are proofreading and you just make all the changes yourself without letting the author see where they have been made, you are missing the opportunity to help develop the author's report-writing skills. In addition, you may inadvertently change the meaning. If the author sees the changes, in effect they are not only learning but also auditing the amendments that the proofreader has proposed. Not a bad idea.

The proofreader is not there to be kind to the author – at least, not in the normal sense. When proposing changes and inserting comments, the proofreader will rarely have the time to be overly courteous, and some people may take offence the first time they get a document back from a proofreader with changes tracked and comments added. In order to reduce the risk of that happening, a proofreader should explain up front that for the sake of speed they will be brief in any comments they make on the document, and that all of the comments should be received on the basis that 'please' and 'thanks' are the sentiments throughout. As I have suggested already, this is an ideal opportunity for you to receive feedback from the proofreader (see Appendix 1, *Feedback: get rich quick*), and it's important for your self-development and your career that both you and they appreciate this. Through doing it this way you get better at producing reports closer to the required standard and, assuming you take on board the feedback, they will need to make fewer corrections the next time.

CAN I SEND IT NOW?

Only once the report has been thoroughly proofread and the final version agreed should you send it out to the customer or intended recipient. In addition, it would be sensible to double-check with your line manager that there are no other reasons (such as an overdue invoice) why the report should not be issued.

Immediately prior to sending out the report, once again ensure that you carry out a final spell check, as well as double-checking areas such as page numbers in the contents list. If you haven't done so already, now is the time to compress any pictures that are within the report. This will help reduce the size of the document and make it easier to send as an attachment to an email, for example. Finally, unless you have specifically agreed otherwise with the customer (e.g. it's a draft document that they may wish to amend) it is wise to convert the document to a PDF. This will ensure that nothing in the document is accidently changed or deleted. A point to watch out for when creating a PDF, is that it is always worth checking symbols and the like (e.g. on maps) after the document has been through this process, to ensure that they still look the same as originally intended.

You will recall that earlier in this chapter I made reference to scenarios where information contained within a report could be of a confidential or sensitive nature. If this is the case, make sure you remind the customer about this when you send them the report. The report may have something included on its title page that makes this clear. Alternatively there may be some additional detail, not within the report, that you need to provide the customer with separately by email.

Some clients will expect to see a draft of the report before you prepare the final version. This is acceptable, and you must be ready to receive and respond quickly to any changes they propose. Again, this is a feedback situation (see Appendix 1), and you should react as you would normally. You must, however, be careful in accepting suggestions from the customer that you do not misrepresent your own findings or conclusions. At the end of the day it is your report, and it is your integrity and credibility that will be at risk if you are persuaded to agree to something that you shouldn't. There can be a fine line between rewording something to give a more accurate or balanced message and changing the meaning to give a completely different impression.

Other customers will not ask or expect to see a document that is labelled 'draft', but when you send a report out to a customer you should always bear in mind, whatever you have called it, that you should treat it as if it were a draft. Be aware that the client may very well come back to you with amendments. If the customer does ask for amendments at this stage you should be as receptive as you would be to any customer requesting to see a draft document in the first instance.

REFERENCES

CIEEM. (2013). *Guidelines for Preliminary Ecological Appraisal*. Chartered Institute of Ecology and Environmental Management. Accessed at: www.cieem.net.

CIEEM. (2015). *Guidelines for Ecological Report Writing*. Chartered Institute of Ecology and Environmental Management. Accessed at: www.cieem.net.

Covey, S. R. (1989). *The 7 Habits of Highly Effective People*. London: Simon & Schuster.

Covey, S. R. (2009): Emotional bank account. Accessed at: www.stephencovey.com/blog/?tag=emotional-bank-account.

Drucker, P. F. (1954). *The Practice of Management*. New York, NY: Harper Business.

Gordon Training International. (undated). Learning a new skill is easier said than done. Accessed at: www.gordontraining.com/free-workplace-articles/learning-a-new-skill-is-easier-said-than-done.

Hundt, L. (2012). *Bat Surveys: Good Practice Guidelines*, 2nd edition. London: Bat Conservation Trust.

IEEM. (2006). *Guidelines for Ecological Impact Assessment in the United Kingdom*. Institute of Ecology and Environmental Management. Accessed at: www.cieem.net.

Mehrabian, A. (1971). *Silent Messages*. Belmont, CA: Wadsworth.

Mehrabian, A. (1972). *Nonverbal Communication*. Chicago, IL: Aldine-Atherton.

Pease, A. and Pease, B. (2004). *The Definitive Book of Body Language*. London: Orion.

Searle, S. M. (2011). *How to Become an Ecological Consultant*. Acorn Ecology.

Appendix 1
FEEDBACK: GET RICH QUICK

The whole subject of feedback cuts across many of the topics covered in this book, and it therefore makes sense to tackle it generically. During the course of your business activities you are going to get feedback. It may come from your line manager, a team member or even a customer. It may not necessarily be called 'feedback' when you are receiving it, but however it is badged you will be well aware of it when it happens.

Feedback is probably one of the most valuable things that you are going to come across within your working environment. It tells you what you are getting right, and as importantly (arguably more importantly) it highlights where you can adopt a different approach or make improvements. You can choose to select the feedback you like and ignore the feedback you dislike, but in doing so you are potentially denying yourself the opportunity to improve and be the best version of yourself that you can be. To be clear, anyone who does not at least listen to feedback and consider what is being suggested is an idiot. Anyone who reacts badly whilst receiving feedback (whether they agree with it or not) is an idiot. And guess what? Anyone who never makes any changes as a result of the feedback they are given? Yes, you get the picture. And someone who does all of these things is the biggest idiot of all. There I go, sitting on the fence once again.

Quite frankly no one is so good at every aspect of everything they do, that someone else can't give them some advice and assistance along the way based on what they have witnessed or experienced. It sometimes takes a bit of courage to give someone constructive feedback as to how they could do things more effectively in the future. Imagine how the person giving the feedback then feels if the recipient reacts badly and 'kicks back'. Most of us would probably then take the approach that we will just not bother to put ourselves in that firing line again. Who loses out? Mr Kickback. He no longer gets any thoughts or advice on how he can develop or improve. At best, he potentially has been the author of his own unfulfilled destiny. At worst, if he doesn't know he is making mistakes because no one feels they can tell him, he could eventually find himself on a performance improvement plan (PIP). And all of this occurs as a result of reacting badly to someone who is trying to help.

Respond well to feedback

I recall one individual working for us who regularly finished off our monthly 121s by asking me, *'Neil, what can I do in order to be better?'* This was highly effective. Not only did that person react well to feedback, they went looking for it. They weren't seeking a pat on the back, they were giving themselves the opportunity to learn and develop along the right lines as quickly as possible. Every time I gave them positive feedback, they said thank you. Every time I gave them areas to work on, again they said thank you. At no time did they ever make me feel awkward about discussing their performance, and this resulted in productive sessions, conducted in a friendly manner on both sides, and with positive intent.

The best approach to take whilst receiving feedback is to listen, say thank you, and ask for more. The more you get, the quicker you will evolve. Table A1.1 provides some further guidance in the delivering and receiving of feedback.

Table A1.1 Delivering and receiving feedback

When delivering feedback ...	When receiving feedback ...
... Give positive as well as corrective feedback	... Stay positive
... Don't shy away from delivering it. You are helping the recipient	... It's an opportunity to learn
... Deliver it quickly (< 24 hours) after the situation being referred to. Don't let it drift	... Listen carefully
... Adopt a friendly manner with positive intent	... Don't react negatively in any way
... Choose the right time to do it (or rather, definitely don't choose a bad time)	... Don't do anything that reduces the chances of you getting more feedback
... Deliver it in private	... Don't be defensive or make excuses
	... Say thanks during and at the end of the session
	... Keep asking for more

So there you have it. You can 'kick back', make excuses and refuse to take on board the feedback that is being given, or you can listen and evolve. As each penny drops and you choose to pick it up, so often you will find it's made of gold.

Appendix 2
EFFECTIVE ALLOCATION OF TASKS

A huge part of a successful working environment lies in the effective allocation and completion of the specific tasks that collectively contribute towards achieving objectives. It is so crucial that time is spent doing the right things, in the right order, to the correct standard, and all of this on time. Task allocation is linked to many of the points raised elsewhere in this book, and is a very important topic in its own right.

A good approach to consider, when allocating tasks, is to take the time to explain *what* needs to be done, *when* it has to be completed, and *who* is responsible. Conversely, when receiving tasks, you should always ensure that the person passing the job to you not only explains what they want, but also gives you a specific deadline. In this scenario the *who* should be obvious, because you are the one who is being asked to do it. It is most definitely not helpful for you to be given a task without a deadline, or with a deadline phrased in such a way that it could be interpreted differently by those involved (e.g. *'it's quite urgent'*).

Referring back to the *what* part of the request, it may also be beneficial to expand upon *why* the task is needed and *how* it should be completed. Table A2.1 provides more detail.

Adopting an approach like this helps ensure that the person receiving the message understands fully what is expected, the relative importance of what they are being asked to do, and the deadline. It all seems pretty clear and concise, doesn't it? However, even when adopting this approach it still creates the opportunity for misunderstandings or mistakes to occur. How can you be sure that the person you are talking to has actually fully taken in what you have told them to do? Also, if you are the person being asked, how can you test that what you think you have heard is actually what was meant? Case study A2.1 gives an example that we can discuss further.

Table A2.1 Model for effective task allocation

What and why	How	When	Who
What needs to happen and why?	Specific guidance as to how to do it	What is the deadline?	Who is responsible?
Produce an internal report on the ecological constraints that may be relevant to this development. It's really important that I have this produced in such a way that it can be used as a discussion document during the meeting we are having with the client on Friday afternoon.	Create the output that shows everything numbered on a site plan, with any relevant buffers applicable. Use a target note approach with an accompanying spreadsheet giving more detail behind each of the findings.	If you can produce a draft by 4 pm on Thursday that will give me a chance to have a look over it in advance of Friday. I can then feedback to you any suggested amendments for you to do on Friday morning. But to be clear, I need the final version before midday on Friday.	John (ecologist)

Case study A2.1   Allocation and receiving of task instructions

Setting: Jane needs some documentation produced ahead of a client meeting later in the week. She is going to ask Robert to do it.

Jane: *Good morning Robert. How are you today?*

Robert: *I am very well, Jane. How are you?*

Jane: *Busy, busy, as always. The meeting with the client went well and a number of the potential issues we perceived can be worked around without any impact on their plans. The otter holt, however, is an issue. Their access bridge can't be moved and the holt, unfortunately, is going to need to be destroyed. I have given them a brief overview as to the licence application considerations and they fully understand what is involved in that respect. But they have asked if we could produce a draft method statement for the destruction of the holt including mitigation, timing of works and the likely costs for our involvement. I would like you to pull all of that together, please.*

Robert: *Of course. I can do that. Jane, would you mind if I confirm back to you what it is that I have been asked to do?*

Jane: *Of course. Please do.*

Robert: *You are looking for a method statement that covers the mitigation plan, including likely timing of works and our associated costs. Is that correct?*

Jane: *Yes, but I have just realised that by using the word 'mitigation' I was meaning for you to also consider compensation for the loss of the holt. My fault. I apologise.*

I have also just realised that I haven't discussed how I want it presented. I would want it in the format of a written document, with the action plan laid out clearly in table format, chronologically. Also, we need a site map showing the location of the existing holt, with the current 'no disturbance' buffer, the location of the new bridge and the location of where the proposed compensation is going to be situated.

Robert: *Perfect. I can do all of that. When do you need this by?*

Jane: *Good question. If I could have a first draft by 3 pm on Wednesday that would fit in well with giving me time to get back to you with changes, so the final version can be ready to go to the client before Friday. I promised that they would have it by the end of the week.*

Robert: *Brilliant, I can do that. I can see you have a lot on your plate today. I'll quickly send you an email confirming what you have asked me to do, and then I'll get things moving later this afternoon.*

Jane: *Thanks for that. That saves me the time writing to you and also ensures that what is getting done is going to be fit for purpose and on time. I look forward to hearing from you and seeing the draft method statement on Wednesday.*

Robert: *You will have it no later than 3 pm Wednesday.*

Jane: *Thanks. Must crack on. Lots to do.*

When Jane hears back what she has requested she realises the obvious has potentially been missed and is able to quickly remedy the situation. The key learning from Case study A2.1 is that, if you are the person receiving the task instruction, it is a really good idea to repeat to the person instructing you what they have asked you to do. Conversely if you are the person asking for something to be done, you should ask the recipient to repeat back to you what they think you have asked of them. In the context of your normal office environment and your usual team mates this is fairly easy to implement, provided of course that everyone has had guidance as to what to do and why it's a good method of working.

Of course, taking the case study as an example, Robert may still fail to produce the goods. If he does fail, Jane can at least be satisfied that the instructions were clear and understood at the time. Better still, to be truly effective Jane would have gone back to her desk and dropped Robert an email documenting precisely what they had agreed and also confirming the deadline. In this instance Robert saved her the time in doing this by saying he would send her an email covering the task request. Whoever takes the initiative, however, putting it in writing is essential at all times. There is no excuse and you can never tell what, many days later, will be genuinely forgotten or tactically denied, resulting in you being made to look ineffective. The best way to ensure that there is never any misunderstanding of communication is to follow it up in writing.

Bringing it all together

Having now considered the effective allocation of tasks, we can also incorporate the wider aspects of what should be happening, including the use of SMART target-setting principles and the application of the RACI model (Chapter 7, *Project planning*), and finding the time to carry out those valuable reviews in order to learn and evolve (Chapter 7, *Review*). Figure A2.1 brings it all together.

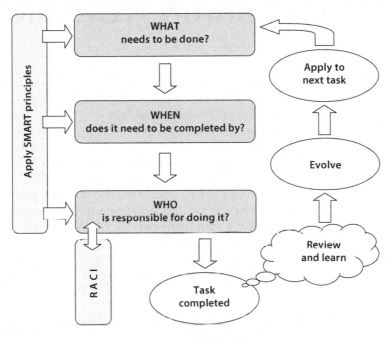

Figure A2.1 Task allocation in conjunction with SMART, RACI and reviews

Appendix 3
EFFECTIVE BRAINSTORMING

Brainstorming may be required in a number of scenarios that an ecologist is likely to encounter. In order to carry out an effective brainstorming session you need to create an environment where as many ideas as possible, relating to the subject being discussed, are forthcoming. The objective is to enable creative thinking, so as to ensure everything is being considered. The more ideas that are put forward in the first place, the more there is to take account of and the more likely it is that any outcomes arrived at are going to be the most appropriate way forward.

There are approaches you can take to help ensure the best outcome is achieved as a result of any brainstorming sessions you are facilitating or participating in. The flipchart shown in Figure A3.1 gives some guidance.

Once the brainstorming session is over the information gathered can then be taken forward to the next stage of the process, which is to discuss the ideas and decide upon the best course of action. Part, or all, of this next stage may occur at a different time and, if appropriate, with different or additional participants.

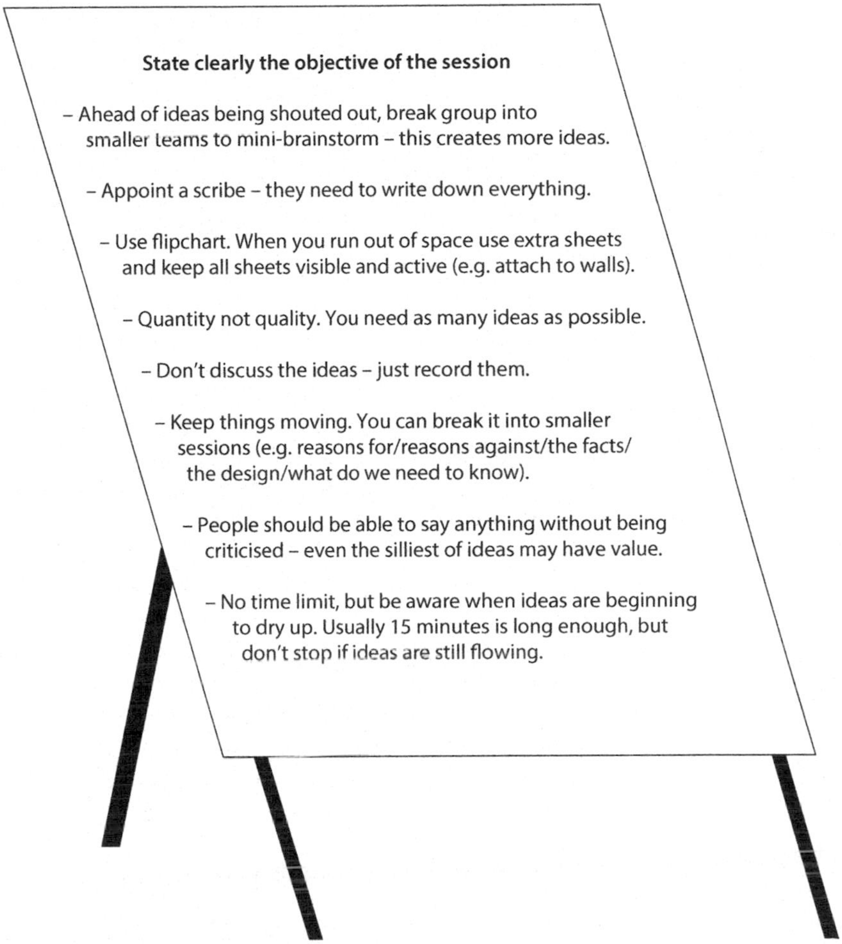

State clearly the objective of the session

– Ahead of ideas being shouted out, break group into
  smaller teams to mini-brainstorm – this creates more ideas.

  – Appoint a scribe – they need to write down everything.

    – Use flipchart. When you run out of space use extra sheets
      and keep all sheets visible and active (e.g. attach to walls).

      – Quantity not quality. You need as many ideas as possible.

        – Don't discuss the ideas – just record them.

          – Keep things moving. You can break it into smaller
            sessions (e.g. reasons for/reasons against/the facts/
            the design/what do we need to know).

            – People should be able to say anything without being
              criticised – even the silliest of ideas may have value.

              – No time limit, but be aware when ideas are beginning
                to dry up. Usually 15 minutes is long enough, but
                don't stop if ideas are still flowing.

Figure A3.1 Facilitating effective brainstorming sessions

GLOSSARY

This is a glossary of business-related words and expressions. It is not a conventional glossary, in that it is not limited to defining words that occur within the book, but is more geared towards terms that you may encounter during your business-related activities. Please note that the words appearing in bold within the definition paragraphs also have their own definitions elsewhere within the glossary

Acting up	This occurs when someone is formally asked to perform at a higher grade, usually for a temporary or predetermined length of time (e.g. whilst their boss is on holiday). Usually it does not mean that the person acting up will get paid more. It is usually more a matter of allowing the business to see how that person performs at the higher level. The benefit for the person acting up, in the main, is that they have experiences and self-development that they might not otherwise have been exposed to, and this could put them in a strong position in the future should a **promotion** opportunity arise.
Annual general meeting (AGM)	A meeting held once per year whereby the owners (e.g. **shareholders** or **partners**) of a business get together in order to discuss various matters relating to the business, including company accounts.
Appraisal	A periodic formal assessment of your performance within the organisation. Often you will be measured against your broadly defined role description along with specific objectives for the period in question. Other, more generic aspects of your work-related performance would also be discussed (e.g. absence record, time keeping).
Associate	An associate is usually someone who works alongside or in conjunction with another person or business, but at a slightly lower status. An associate would not normally be directly employed by the business.

Associated company	An associated company is one which is associated with another company at a point in time, and this occurs if one of the companies has control over the other one or both companies are controlled by the same person or group of people. Rules are in place in order to determine when companies are legally regarded as being associated to each other.
Balance sheet	The section within a set of company accounts that shows the totals of what is owned by the company against what is owed by the company.
Blue line boundary	Appears on maps and plans relating to planning applications. Used in conjunction with the **red line boundary**. The blue line indicates all other land owned by the applicant adjacent to or close to the land that is subject to the proposed development (outlined in red).
Bonus	Extra money paid to you by your employer, for example, as a result of you exceeding targets or working particularly hard on a project, or the business performing well for the period in question. Paid over and above your normal salary, but still subject to tax, national insurance, etc.
Cashflow	No matter how profitable a business is, it cannot survive without actual cash in the bank. **Profit** and cash are two totally different ways of looking at a business. Many profitable businesses fail because they don't have enough cash to pay creditors. When you hear about a business having cashflow problems it doesn't necessarily mean they are not profitable, it may mean that they are spending lots of money or don't have much money coming into the business. They may be awaiting the payment of **invoices** due to them. The important thing is that business owners constantly monitor and manage their cashflow so that they don't spend money at the wrong time, and therefore risk running out of cash.
Company accounts	Formal accounts prepared, usually annually, by an accountant in order for information to be given to HM Revenue and Customs, and in the case of **limited companies** and **limited liability partnerships**, Companies House.

Constraints map In general terms this is a map of an area within which various boundaries and buffers are added in order to highlight that beyond or within those areas restrictions in activities may or would be applicable. It is often used to conversely establish the areas where there are no constraints and therefore these areas are easier to work within, for example when considering the design of development plans and the like.

Contract fee When working on a contract fee basis, the person contracted to do the work will be paid to complete the contract as required irrespective of how long it takes. There is a risk that the contract may take longer than envisaged and the contractor may lose out, whereas on the other hand it may be completed ahead of schedule, to the benefit of the contractor. This is a different method of assessing what will be paid at the end of a contract, when compared to a **day rate** arrangement.

Corporate brochure A marketing brochure or booklet that is used to help promote the business in a positive light. Will often provide details such as the history of the company, its resources, its products and strengths. May cover areas such as vision and mission statements. May be designed in such a way as to attract a certain type of customer. There may be different versions in order that a more tailored approach can be conveyed to different customer types/sectors.

Day rate When working on a day rate then the person contracted to do the work will be paid according to how long the work has taken to be completed. This is a different method of assessing what will be paid at the end of a contract, as compared to a **contract fee** arrangement.

Deadline A date or time by which something must be completed, in accordance with what was instructed and to the required standard. Not to be missed.

Director Only applies to **limited companies**. Usually someone who has been appointed by the **shareholders** of a limited company in order to manage the company as a whole (managing director) or is responsible for specific aspects of the company's performance (e.g. finance director). Note that a director may themselves also be a shareholder of the business. In many smaller companies it would not be unusual for the directors and the shareholders to be the same people.

Expenses	Money claimed back by an employee when they have used their own personal money to purchase items or services on behalf of the company. May refer to fuel, travel costs, postage, accommodation and the like. Expenses are paid back in full without deductions and are not subject to national insurance or tax deductions.
Fee proposal	A **quote** that outlines how much the work is anticipated to cost, and the manner and the timescale within which it will be carried out. Would usually also contain other relevant details as requested from the potential customer (e.g. details of experience, health and safety documentation, insurance confirmation, corporate brochure). See also **tender**.
Financial year end (FYE)	The date when a business's financial accounts are made up to, occurring at the end of each financial year. Although not a necessity, more often than not an FYE would be at the end of a calendar quarter (31 December, 31 March, 30 June or 30 September). Many business owners can be quite busy in the period leading up to and immediately after their FYE, as they need to gather lots of information together at that point to pass over to their accountants.
Freelance, freelancer	Self-employed person who works alone and is available for hire for the skillset(s) they are able to deliver. Similar to a **sole trader**, but you would not expect a freelancer to have staff directly employed.
Gardening leave	This is when an employee is leaving a job (having resigned or had their employment terminated) and is told to not come back to the workplace during their notice period. During this period they will still be paid as normal, but despite not working for that employer they will not be allowed to carry out work for anyone else until such time as they have officially left the business. Gardening leave is used in a variety of circumstances, for example to prevent employees from taking sensitive information with them before leaving to work for a competitor. It may also be appropriate if the employee has been made **redundant** and hence their job no longer exists.
Geographic Information System (GIS)	A GIS is used in order to create, analyse, interpret and present information relating to spatial or geographic data (e.g. maps). GIS software programmes are available in order to allow this to happen.

Invoice	In order to be paid for work carried out, a company needs to raise an invoice to a customer. Each invoice would be issued in accordance with the company's template, and each invoice should show a unique identification number. In addition, if the company is VAT registered, then VAT details should be shown.
Joining instructions	The information you are sent in advance of attending a training course or a conference. These would typically include details of the location of the event, the timing, the programme, other delegates attending and where to park.
Limited company (Ltd)	A limited company is a legal entity in its own right. It is owned by **shareholders** whose personal liability is restricted to the value of their shareholdings. These shareholders may be individuals or other companies. The business has to be registered with Companies House, and must also submit returns along with copies of their annual accounts to Companies House every year. Normally the accounts must be audited. A limited company must have at least one **director**. Limited companies pay corporation tax based on their **profits**. Company directors are employees and therefore pay personal tax and national insurance contributions in the same way as any other employed person.
Limited liability partnership (LLP)	These are businesses which are set up in a similar way to a **partnership** (i.e. the business is owned by **partners** as opposed to shareholders) but with the partners' personal liability limited to the amount of money they have invested in the business (i.e. similar to the limited liability status that shareholders have in a private **limited company**). Unlike a conventional partnership, LLPs must register with and file accounts with Companies House in a similar manner to that required by private limited companies.
Management information (MI)	Information about the performance of the business that is easily accessible to those in a decision-making capacity (e.g. managers) within the business. May relate to any or all financial and performance aspects of the business at any given time or over any particular period. May also be used to make judgements about key decisions and anticipated business performance.

Mentor
Someone who is asked or has offered to be a point of referral or a source of guidance in order to help you develop in a particular direction. The person acting as a mentor would not normally be paid to do this, and indeed may not even work within the same business. Usually, for mentoring to be effective, the mentor would be someone that is much further advanced in their career or has more experience than you (e.g. someone you would aspire to be, in terms of their career and/or performance levels).

One-to-one (121)
A periodic meeting with your line manager when various matters relating to your performance, development needs, working conditions and specific cases are discussed. The meeting is not a performance review in the same way as an **appraisal**, and the employee should be given the opportunity to add things to the agenda that they wish to talk about.

Overtime
Time officially worked beyond the normal expectations or terms of your contract of employment. Usually, when the term overtime is used you would expect to receive additional salary to reflect the extra hours worked, this additional salary being subject to national insurance and tax. In some circumstances you may be required to work overtime and not be paid for it, but be given the time back as **time off in lieu**. This is not uncommon within the ecology sector, bearing in mind the seasonal variation in workloads that often apply.

Partner
One of the owners (there must always be at least two) of a **partnership**. A partner will own a percentage of the business.

Partnership
A legal term for a business structure whereby the owners of the business are described as **partners** and each will own a proportion of the business. Usually there will be a partnership agreement in place. Each partner is regarded as self-employed for national insurance and tax purposes. Unlike a private **limited company**, the partnership itself has no legal status distinct from the partners. A partnership does not file accounts with Companies House and is not governed by the same rules that would apply to a private limited company. One way to view a partnership is to think of it as at least two self-employed people coming together to form a business. It's all the same as this, other than in name and with a partnership agreement being in place.

Payment terms	The number of days elapsing after the issue date appearing on an **invoice** within which the company issuing the invoice expects to be paid.
Performance improvement plan (PIP)	Plan put in place by a line manager in order to help an employee improve aspects of their performance that are currently below acceptable.
Performance review	See **appraisal**.
Personal protective equipment (PPE)	This refers to items that are worn by an employee during the course of their work in order to help keep them safe. It includes high-visibility clothing, safety boots, hard hat, gloves, eye protection and ear protection.
Probation period	A 'trial' period applying when commencing a new job in a different business, where normal rules regarding disciplinary procedures and the like do not apply. Salary or other terms and conditions may also be impacted upon during this period, with a view to these being improved once the probationary period comes to an end. In effect a probation period allows the employer to assess that you are capable of performing at the level required for the role.
Profit	If a business isn't profitable then this raises lots of questions about its viability in the longer term. Some businesses can afford to have bad years in this respect, with the shortfall more than made up in good years. Profitability does not, however, equate to cash in the bank (see **cashflow**), and quite often it is businesses that, irrespective of their short-term profit margins, do not manage their cashflow properly that struggle to survive.
Promotion	When you are given an increase in responsibility at a higher position within the business, with more salary and/or better terms and conditions of employment applying.
Public limited company (PLC)	A PLC is similar to a private **limited company** but must have at least two **shareholders** and at least two **directors**. It has the ability to issue shares to the general public. Usually PLCs are listed on the Stock Exchange. The rules governing PLCs are much tighter and cover a number of matters that do not apply to a private limited company.

Purchase order (PO)	Not always applicable, but for many organisations (the larger the more likely to be the case) their Accounts Payable department would require a PO before an invoice is paid. The PO in effect legitimises the **invoice** and confirms that someone in the business has properly authorised the expenditure, provided of course that the product or service has been delivered as required and to the correct specification (e.g. fit for purpose).
Quote	See **fee proposal** and **tender**.
Red line boundary	Appears on maps and plans relating to planning applications. Used to show the area that is subject to the proposed development, including access points and the like. There may also be an additional **blue line boundary** if the developer owns other land adjacent to or close to the land that is subject to the proposed development.
Redundancy	Payment made to you (as per the law) as a result of your employer no longer being able to keep you on, usually due to cutbacks or the role you are carrying out no longer existing. Note that you would need to have worked for the same employer for a specific period of time before you would be entitled to anything. Where this does apply, there are legal minimum payments.
Secondment	This usually means that you are transferred to another department or location within the same business for a set period of time. It may be to make up for a shortage in staffing elsewhere, or to assist with a particular project, or to have your particular expertise attached to a location or project. This is not the same as **acting up** or a **promotion** in that you would not normally be operating at a higher level, and would not be receiving better salary or terms and conditions whilst on secondment. Things that may apply could be an agreement to pay additional incurred **expenses** (e.g. rent of a house, weekend travel home) and the payment of **subsistence** whilst away on secondment.
Shareholder	A person (or another organisation) who owns a shareholding within a private **limited company** or **public limited company**. In effect the shareholders collectively own these types of business. Shareholders may also be employed within the business, although there is no legal requirement for this to be the case.

Sole trader
Someone who is registered with the tax office as self-employed. They alone own the business, which is not a **limited company**, a **partnership** or **limited liability partnership** (i.e. there are no **partners** or **directors**). They may still have employees. They do not lodge accounts with Companies House, but they have to pay national insurance contributions and make an annual self-assessment tax return to HM Revenue and Customs.

Subcontractor
An individual (e.g. a **freelancer**) or another business employed by a company to carry out work on their behalf in return for payment. Subcontractors are not direct employees of the company employing them, but from an insurance point of view, should something go wrong, and a subcontractor is negligent whilst working for the company, the company itself may in the first instance have to deal with any claims. Likewise if a subcontractor is injured during the work it is possible that a claim will be made against the company's employer's liability insurance policy. Usually, when subcontractors are employed, legal agreements are put in place, in addition to which the company employing them runs checks on their insurance policies, health and safety procedures, etc.

Subsidiary company
A subsidiary company is one which is wholly owned or controlled by another company. The controlling company is called the 'parent company'. Rules are in place to determine when companies are legally regarded as being 'subsidiary' or 'parent' to each other.

Subsistence
An amount you receive for working away from home for an extended period of time (e.g. overnight with requirement that you need to stay in a hotel). The amount would normally be limited to a fixed figure acceptable to HM Revenue and Customs in order to ensure that it cannot be construed as a benefit in kind. It is there to reflect that you may, by the very nature of not being at home, incur additional **expenses** (e.g. having to buy a bar supper and make additional phone calls). These payments would not normally be subject to tax and national insurance.

Supplier
A person or organisation that provides your business with a product or a service.

Tender, tendering   This refers to the process whereby a company is formally seeking to acquire services from another business (e.g. a **subcontractor**). In such circumstances the **subcontractor** is asked to tender for the work, and if so inclined they would put together a tender document (i.e. **fee proposal**) and submit it along with others also tendering for the same piece of work. Eventually they will hear back as to whether or not they have been successful. In some circumstances, depending upon the customer type, there can be strict rules applicable as to how you should behave during a tendering process.

Time off in lieu (TOIL)   Time giving back to an employee as compensation for additional hours worked over and above the normal expectations or terms of their contract of employment. May be given instead of a monetary payment for **overtime**, and is used regularly within the ecology sector, bearing in mind the seasonal variation in workloads that often apply.

INDEX

Page numbers in *italic* indicate figures or tables and in **bold** indicate glossary terms.

121 meetings 22, *126*, 181, **193**

abbreviations, in reports *174*
accountability 143, *144*
acknowledgements 151, *175*
acronyms, in reports *174*
acting up **188**
action plan templates 168, *169*
action points
   meetings 120, 121, 123
   report recommendations
     168–9, *169*
active listening 57–8, *58*, 65
advice, asking for 22, 103–4
agendas, meeting 116–17, *119*, 120,
   124
annual general meetings (AGMs) **188**
answer machines 92
any other business (AOB) *119*
apologies 56–7
appearance *39*, 64
appendices, in reports *159*, 160, 166, 170
appraisals 22, *125*, **188**
assistance, accepting 42
associated companies **189**
associates **188**
attendees, meeting 115–16, *116*, 121–3,
   124
audience, for reports 157–8

bad news, delivering 78, 100–1, 148–9
balance sheets **189**
Barber, Jerry 27
blue line boundaries **189**
body language 50–1, *51*, 52, 63–5, *66*
   matching, mirroring and pacing 65–7,
     67
   and positive active listening 57, *58*, 65
bonuses **189**
brainstorming 146, 166, 186, *187*
breaks, taking 43
brochures, corporate 36, **190**
buffer zones 164, 168

case studies
   closed questions 59–61
   customer satisfaction 7–8
   email techniques 74–5
   managing visitors to desk 88
   minute taking 122
   open questions 61–2
   poor performance 41
   professionalism 30–2
   punctuality 34–5
   task allocation 183–4
   telephone techniques 54–5
   written communications 68–70
cashflow **189**
chairing meetings 118, *118–19*, 120

chairs, guest 87, 88
CIEEM (Chartered Institute of Ecology and Environmental Management) guidance 155, 157, 161, 167
clarity, in reports *175*
close phase of projects 151
closed questions 59–61, *63*
coaching 22, 171, 177
commitment 40
communications 49–50, *49*
　components of 50–2, *51*
　face-to-face *51*, 57, *58*, 74
　listening 57–8, *58*, 65
　with manager 19, 45–6, *46*
　preferences 78
　project close 151
　project progress 148–50, *149*
　questioning 58–62, *63*
　self-awareness 65, 77–80
　tonality 50, *51*, 52
　words in 50, 51–2, *51*
　see also non-verbal communications; phone calls; spoken word; written communications
company 45
　criticising 33, *126*
　maintaining reputation of 33, 45
　researching 35–7
company accounts **189**
company websites 36
competence, four stages of 7
conclusions sections of reports 167–8
confidential information, in reports 166
conflicts, in task prioritisation 105
constraints maps **190**
contact details, in reports 162–3
contacts data, phone 92
contract fees **190**
controllables 20, 22, 27
　see also positive behaviours
copyright 162, 171
corporate brochures 36, **190**
corrective feedback 22, 23, *126*

costs
　meetings 115
　mobile tariffs 93
　recruitment process 17
Covey, Stephen R 49, 105, *105*
credibility 10, 29, 44–5, 53, 94, 151, 178
cultural fit 2
current-year objectives (CYOs) *125*
customer knowledge 132, 135–6
customer meetings *126*
customers' perspectives
　on professional service 7–8, 28–9, *28*
　on reports 156, 157–8, 178
CYOs *see* current-year objectives (CYOs)

Darrell-Lambert, David 158
day rates **190**
dead time, using *98*, 107–9, *108*
deadlines 9, 10, 75–6, **190**
　adding to diary 96
　management of 101–3, *102*
　project 141, *142*, 147, *147*
　report 171–2
　task 182, *183*
delegating tasks 90, *90*, 105, 106–7
desk environment *85*, 86–9
desk studies 165, 166
diary management *47*, *85*, 86, 87, 94–6, *118*
directors **190**
discretion 45
discussion sections of reports 166–7
distractions
　phones 87, 89, 91–2, 94
　reducing 87–9
draft reports 178
dress code *39*
driving, safety 43
Drucker, Peter 1

early days *see* new employees
effectiveness 2–3, 5–7, 8–10
efficiency 8–10

emails 73–7
  chains 75–6
  following up conversations 68, 70–1,
    71, 184
  format 72, 73–5
  inbox organisation 85, 89–91, 90
  meeting invitations 115
  'one touch' approach 90–1
  professionalism 73, 74–5, 76
  'reply all' 69–70
  salutations and sign offs 72, 74
  subject line 76–7
employees
  effective 5
  see also established employees; new
    employees
employees' perspectives 4
  on new job 17–18
  on objectives of job 18, 19–21, 19
  on poor performance 21, 22
  on recruitment process 15, 16
employers
  effective 5–6
  see also company; managers
employers' perspectives 4
  on new team member 17
  on objectives of job 19, 19
  on poor performance 22, 23
  on recruitment process 14–15, 16, 17
employment
  finding first position 13–15
  key objectives of 18–21, 19
  managing poor performance 21–3,
    41
  moving on 21, 33
  recruitment process 2, 6, 14–15, 16, 17
  starting new jobs 15–18
enemies, making 37
enthusiasm 40
established employees
  communicating with managers 45–6,
    46
  credibility 29, 44–5
  enthusiasm 40

hitting objectives 43–4
  positive behaviours 40–7, 47
  working extra hours 40–1
  working long hours 42–3
executive summaries 159, 159, 163,
    164
expenses 191
experience 1
external stakeholders see stakeholder
    management
extra hours, working 40–1

face-to-face communications 51, 57, 58,
    74
fee proposals 191
feedback 180–1, 181
  corrective 22, 23, 126
  project review 152, 185
  from proofreaders 177
feedback sessions 22, 126, 181
fees
  contract 190
  day rates 190
field work
  being prepared for 39, 85
  health and safety 43, 85
  in reports 165, 166
figures, in reports 175
financial year end (FYE) 191
first impressions 35, 38–40, 39
Ford, Henry 84
Franklin, Benjamin 129
Freed, Arthur 13
freelancers 191
friendships 37, 46
FYE see financial year end (FYE)

Gantt charts 146–8, 147, 149
gardening leave 191
Geographic Information Systems (GIS)
    191
gossip 45
gross misconduct 33
guest chairs 87, 88

hairstyles 64
handshaking technique *39*
headers and footers, in reports *174*
health and safety
    driving 43
    field work 43, *85*
    site induction meetings *127*
    working excessive hours 43
holidays 93, 95
home working 89
honesty *47*, 104
human nature 63–5

integrity 8, 30, 45, 151, 178
interruptions
    phones 87, 89, 91–2, 94
    reducing 87–9
interviews 2, 15, 33
introduction sections of reports 163–4
introductions, in meetings *118*, *124*
invoices **192**
IT environment 85, 89–91, *90*

Jackson-Matthews, Steve 2, 13–14, 30,
    76, 171
job security 44
jobs *see* employment
joining instructions **192**

Keller, Helen 113
knowledge, technical 1

lateness *see* punctuality
letter writing 71–3, *72*
limited companies (Ltd) **192**
limited liability partnerships (LLPs) **192**
listening 57–8, *58*, 65
Lombardi, Vince 22
long hours, working 42–3

management information **192**
managers
    communicating with 19, 45–6, *46*
    effective 5–6

gaining support of 44
managing poor performance 22, 23,
    41
relationships with 23, 32–3
supporting 20–1
managers' perspectives 4
    on new team member 17
    on objectives of job 19, *19*
    on poor performance 22, 23
    on recruitment process 14–15, *16*, 17
managers, project 134–5, 143
maps
    constraints **190**
    in reports 168, 173–4, *174*
matching 65–7, *67*
meetings 113–14
    action points 120, 121, 123
    agendas 116–17, *119*, 120, *124*
    annual general **188**
    any other business (AOB) *119*
    attendees 115–16, *116*, 121–3, *124*
    chairing 118, *118–19*, 120
    costs 115
    customer *126*
    formal contributors 123–5
    health and safety site induction *127*
    invitations 115
    location 115, *116*
    meeting rooms 117
    minute taking 120–1, *122*
    one-to-one 22, *126*, 181, **193**
    organising 114–17, *116*
    performance reviews 22, *125*, **188**
    presentations 124–5
    project review 150, 152
    time keeping *118*, *119*, *124*
    timing of 115, *116*
mentors **193**
messages, telephone 53, 54–5, 92, 93, 94
methods sections of reports 164–5
Microsoft Excel 146
milestones
    adding to diary 96
    project *147*, 148, 149

minute taking 120–1, 122
mirroring 65–7, *67*
misconduct 33
mistakes, accepting responsibility for 22, 45, *47*, 53, 56–7
mobile phones *39*, 87, 89, 91–4, *124*
monitoring
  new employees 17
  project progress 148–50, *149*
MS Word 177
multiple task delivery 107–9, *108*

new employees
  asking questions 35–6, *39*
  building relationships 37–8, *39*
  first impressions 35, 38–40, *39*
  fitting into team 2, 37
  monitoring of 17
  positive behaviours 33–40, *39*
  punctuality 34–5, *39*
  report writing 171
  researching the company 35–7
  starting new jobs 15–18
news, delivering bad 78, 100–1, 148–9
newsletters 150
non-verbal communications 50–1, *51*, 52, 63–5, *66*
  matching, mirroring and pacing 65–7, *67*
  and positive active listening 57, *58*, 65
notebooks *47*

objectives
  breaking down project 140–1, *140*
  current-year (CYOs) *125*
  of employment 18–21, *19*
  hitting 43–4
  of reports 156–7
  SMART 145, *145*, *185*
  understanding project 132–3
one-to-one meetings 22, *126*, 181, **193**
open questions 61–2, *63*
opportunities, in task prioritisation 105
organisational charts *39*

organisational skills 83–4
  desk environment *85*, 86–9
  diary management *47*, *85*, 86, 87, 94–6, *118*
  disorganised behaviours *85*
  email inbox organisation *85*, 89–91, *90*
  IT environment *85*, 89–91, *90*
  knowing where things are 85–6, *85*
  managing visitors to desk 87–8
  organisational excellence 109–10
  phones 91–4
  prioritisation 84, 103, 105–7, *105*, *106*
  reducing distractions 87–9
  saying 'no' 109
  taking on additional responsibilities 109–10
  see also time management
overtime **193**

pacing 65–7, *67*
PAL *see* positive active listening (PAL)
partners **193**
partnerships **193**
  limited liability **192**
passion for job 40
payment terms **194**
PDF format for reports 178
performance improvement plans (PIPs) 180, **194**
performance, managing poor 21–3, 41
performance reviews 22, *125*, **188**
personal calls *39*, 89
personal hygiene *39*
personal protective equipment (PPE) *85*, **194**
personality profiling 79–80
perspectives 3–4
  see also customers' perspectives; employees' perspectives; employers' perspectives
phone calls *51*
  delivering bad news 78, 100–1, 148–9
  effective techniques 53–5, 58, *58*

instead of email 73–4
making difficult 53–4, 78, 100–1
messages 53, 54–5, 92, 93, 94
personal *39*, 89
positive active listening 58, *58*
in public places 89
phones
    contacts data 92
    distractions 87, 89, 91–2, 94
    mobile *39*, 87, 89, 91–4, *124*
    one versus two 92–3
    organisation of 91–4
    voicemail 92, 93, 94
pictures/photographs, in reports 161–2,
    162, 168, 173, *173*, *174*, 178
PIPs *see* performance improvement
    plans (PIPs)
planning *see* project planning
plans, in reports 168, 173–4, *174*
PLCs *see* public limited companies
    (PLCs)
POs *see* purchase orders (POs)
positive active listening (PAL) 57–8, *58*,
    65
positive behaviours 20, 27
    accepting responsibility 22, 45, 47, 53,
        56–7
    asking questions 35–6, *39*
    building relationships 37–8, *39*
    communicating with manager 19,
        45–6, *46*
    credibility 10, 29, 44–5, 53, 94, 151,
        178
    early in new role 33–40, *39*
    enthusiasm 40
    established employees 40–7, *47*
    first impressions 35, 38–40, *39*
    hitting objectives 43–4
    new employees 33–40, *39*
    punctuality 34–5, *39*, *118*, *124*
    researching the company 35–7
    supporting manager 20–1
    top ten ongoing *47*
    working extra hours 40–1

working long hours 42–3
    see also professionalism
PPE *see* personal protective equipment
    (PPE)
praise, giving 22
presentations 124–5
prioritisation 84, 103, 105–7, *105*, *106*
probation periods **194**
probing questions 62, *63*
procrastination 99–101
professionalism 27–33, *39*, 151
    appearance *39*, 64
    case study 30–2
    customer service 7–8, 28–9, *28*
    in written communications 73, 74–5,
        76
profits **194**
progress reporting 149–50, *149*
project management 129–32, *131*, *132*
    close phase 151
    customer knowledge 132, 135–6
    defining projects 129, 130
    development cycle *152*
    monitoring progress 148–50, *149*
    progress reporting 149–50, *149*
    review phase *152*, *185*
    stakeholder management 136–40, *137*,
        138, *139*
    teams 133–5
    understanding objectives 132–3
    see also project planning
project managers 134–5, 143
project objectives
    breaking down 140–1, *140*
    understanding 132–3
project planning 140–8
    breaking down objectives 140–1, *140*
    deadlines 141, *142*, 147, *147*
    Gantt charts 146–8, *147*, 149
    milestones *147*, 148, 149
    RACI model 142–4, *144*, *185*
    scheduling tasks 146–8, *147*
    SMART targets and objectives 145,
        145, *185*

task allocation 145, *145*
task identification 141, *142*
project websites 150
projects, defined 129, 130
promotion **194**
proofreading 175–7
public environments, working in 89
public limited companies (PLCs) **194**
punctuality 34–5, *39*, *118*, *124*
purchase orders (POs) **195**

quality versus speed 9, 10
questioning
    effective 58–62, *63*
    in new role 35–6, *39*
questions
    closed 59–61, *63*
    open 61–2, *63*
    probing 62, *63*
quotes *see* fee proposals; tenders

RACI model 142–4, *144*, *185*
raw data, in reports 160
recommendations sections of reports
    167, 168–9, *169*
recruitment process 2, 6, 14–15, *16*, 17
red line boundaries **195**
redundancy **195**
references sections of reports 169–70
relationships
    building 37–8, *39*
    with manager 23, 32–3
reminders 86, 91, 96
'reply all' to emails 69–70
report writing 155–6
    action plan templates 168, *169*
    appendices *159*, 160, 166, 170
    audience 157–8
    conclusions sections 167–8
    contact details 162–3
    desk study data 165, 166
    discussion sections 166–7
    executive summaries 159, *159*, 163,
        164

field work data 165, 166
introduction sections 163–4
maps and plans 168, 173–4, *174*
methods sections 164–5
objective and purpose 156–7
pictures/photographs 161–2, *162*,
    173, *173*, *174*, 178
practical advice 170–4, *174–5*
proofreading 175–7
recommendations sections 167, 168–9,
    169
references sections 169–70
report structure 159–60, *159*, 161, *161*
results sections 165–6
sending out reports 177–8
sensitive data 166
site details 164, 173–4, *173*
table of contents 163
title pages 161–2, *162*
writing style 158, *174*
reporting, project progress 149–50, *149*
reports
    project completion 151
    see also report writing
requirements, in reports 168
research, in new role 35–7
responsibilities
    project 143, *144*
    taking on additional 109–10
responsibility, accepting 22, 45, *47*, 53,
    56–7
results sections of reports 165–6
review meetings
    performance 22, *125*, **188**
    project 150, 152
review phase of projects 152, *185*
Rohn, James 101
Roosevelt, Theodore 83

salutations, in written communications
    72, 74
Searle, Susan 14
secondment **195**
secrets 45

self-awareness 65, 77–80
self-development 152, *152*
sensitive data, in reports 166
shareholders **195**
signature lines, in written
    communications *72*, 74
Singleton, Reuben 22, 73, 156
site details, in reports 164, 173–4, *173*
SMART targets and objectives 145, *145*,
    185
social media 36, 87, 94, 150
software packages 91, 146, 177
sole traders **196**
sorry 56–7
speed versus quality 9, 10
spell checking 176, 178
spoken word 52–7
    apologies 56–7
    following up conversations in writing
        68, 70–1, *71*, 184
    message taking 54–5
    telephone techniques 53–5, 58, *58*
    thanks 56
    tonality 50, *51*, 52
    see also phone calls
staff development 6
stakeholder management 136–40, *137*,
    138, *139*
    RACI model 142–4, *144*, *185*
    understanding customers 132, 135–6
stalling 103–4
subconscious behaviour 7, *7*, 67
subcontractors **196**
subject line, email 76–7
subsidiary companies **196**
subsistence **196**
suppliers **196**
surgery approach to queries 88
symbols, in reports *174*

table of contents, in reports 163
tables, in reports *175*
targets
    hitting 43–4

setting 102, *102*
SMART 145, *145*, *185*
see also deadlines
task allocation
    delegating 90, *90*, 105, 106–7
    effective 182–5, *183*, *185*
    meeting action points 120, 121
    project tasks 145, *145*
    report recommendations 168–9, *169*
task management flow chart *145*
tasks, project
    allocating 145, *145*
    deadlines 141, *142*, 147, *147*
    identifying 141, *142*
    scheduling 146–8, *147*
    see also action points
TBT *see* toolbox talks (TBT)
team development 6
teams
    building relationships 37–8, *39*
    fitting into 2, 37
    project management 133–5
    technical knowledge 1
technology
    in meetings *119*
    see also emails; mobile phones
telephone techniques 53–5, 58, *58*
    see also phone calls
templates
    action plan 168, *169*
    report 170, 171, 172
    for written communications 71
tenders **197**
thanks 56
time creation 107–9, *108*
time keeping
    meetings *118*, *119*, *124*
    punctuality 34–5, *39*, *118*, *124*
    working extra hours 40–1
    working long hours 42–3
time management 84, 97–104
    asking for advice 103–4
    deadline management 101–3, *102*
    multiple task delivery 107–9, *108*

procrastination 99–101
report writing 171–2
time-saving behaviours 97–8, *97, 98*
time-wasting behaviours 84, *85,* 86,
   97–8, *97, 98*
using dead time *98,* 107–9, *108*
time off in lieu (TOIL) **197**
time sheets 104
tiredness 43
title pages of reports 161–2, *162*
'to do' lists 96
TOIL *see* time off in lieu (TOIL)
tonality 50, *51,* 52
toolbox talks (TBT) 168, *169*
traffic light reporting *149,* 150

visitors to desk 87–8
visual behaviours *see* non-verbal
   communications
voice-to-voice communications *see*
   phone calls
voicemail 92, 93, 94
volunteering 109–10

websites
   company 36
   project 150
Wilde, Oscar 155
words, in communications 50, 51–2, *51*
work-related experience 1
working day/week
   typical 14

working extra hours 40–1
working long hours 42–3
working environment 84
   desk environment *85,* 86–9
   diary management *47, 85,* 86, 87,
      94–6, *118*
   disorganised behaviours *85*
   email inbox organisation *85,* 89–91,
      90
   IT environment *85,* 89–91, *90*
   knowing where things are 85–6, *85*
   managing visitors to desk 87–8
   phones 91–4
   reducing distractions 87–9
written communications *51,* 68–77
   email chains 75–6
   email inbox organisation *85,* 89–91,
      90
   email techniques *72,* 73–7
   following up conversations 68, 70–1,
      71, 184
   letter writing 71–3, *72*
   meeting invitations 115
   problems with 68–70, 73
   professionalism 73, 74–5, 76
   'reply all' to emails 69–70
   salutations and sign offs *72,* 74
   templates 71
   see also report writing

Young, Edward 99